机械工程专业英语

（第 16 版）

施平　主编

哈尔滨工业大学出版社

内 容 提 要

本书以培养学生专业英语能力为主要目标。内容包括:力学、机械零件与机构、机械设计、机械制造、管理、现代制造技术、科技写作。全书共有 62 篇课文,其中 33 篇课文有参考译文。

本书可以作为机械设计制造及其自动化,机械工程及自动化,机电工程等专业的专业英语教材,也可以供从事机械工程专业的科技人员参考使用。

图书在版编目(CIP)数据

机械工程专业英语/施平主编.—16 版.—哈尔滨:哈尔滨工业大学出版社,2015.3
ISBN 978-7-5603-5246-6

Ⅰ.①机… Ⅱ.①施… Ⅲ.①机械工程-英语-高等学校-教材 Ⅳ.①H31

中国版本图书馆 CIP 数据核字(2015)第 032507 号

责任编辑	田 秋 孙 杰
封面设计	卞秉利
出版发行	哈尔滨工业大学出版社
社　　址	哈尔滨市南岗区复华四道街 10 号　邮编 150006
传　　真	0451-86414749
网　　址	http://hitpress.hit.edu.cn
印　　刷	哈尔滨市石桥印务有限公司
开　　本	880mm×1230mm　1/32　印张 11.5　字数 390 千字
版　　次	2015 年 3 月第 16 版　2015 年 3 月第 1 次印刷
书　　号	ISBN 978-7-5603-5246-6
定　　价	26.00 元

(如因印装质量问题影响阅读,我社负责调换)

再版前言

专业英语是大学英语教学的一个重要组成部分,是促进学生们完成从英语学习过渡到实际应用的有效途径。教育部颁布的"大学英语教学大纲"明确规定专业英语为必修课程,要求通过四年不断线的大学英语学习,培养学生以英语为工具交流信息的能力。根据此精神编写了这本《机械工程专业英语》教材,以满足高等院校机械工程各专业学生们的专业英语教学需求。

本书所涉及的内容包括:力学、机械零件与机构、机械设计、机械制造、管理、现代制造技术、科技写作等方面。通过这本教材,学生们不仅可以熟悉和掌握本专业常用的及与本专业有关的英语单词、词组及其用法,而且可以深化本专业的知识,从而为今后的学习和工作打下良好的基础。

在此次再版前,编者们吸取了多所大学在使用本书过程中提出的许多宝贵意见,对全书进行了修订和补充。全书由62篇课文组成,其中33篇课文有参考译文。本书选材广泛,内容丰富,语言规范,难度适中,便于自学。

本书由施平主编,参加编写工作的有胡明、乔世坤、田锐、施晓东、魏思欣、李越、侯双明、王旭、张宏祥、杜勇、孙德金、王海军、郭启臣、胡淼、陶文成、梅竹、丁印成、郭晓江、林晓东、张志伟、梁东伟,由贾艳敏担任主审。由于水平有限,书中难免有不足和欠妥之处,恳请广大读者批评指正。

编 者
2015年3月

Contents

1 Basic Concepts in Mechanics ……………………………………… (1)
2 Simple Stress and Strain ………………………………………… (5)
3 Forces and Moments ……………………………………………… (9)
4 Overview of Engineering Mechanics …………………………… (12)
5 Shafts and Couplings …………………………………………… (15)
6 Shaft Design ……………………………………………………… (19)
7 Fasteners and Springs …………………………………………… (24)
8 Belt Drives and Chain Drives …………………………………… (28)
9 Strength of Mechanical Elements ……………………………… (33)
10 Physical Properties of Materials ……………………………… (38)
11 Rolling Bearings ………………………………………………… (42)
12 Spindle Bearings ………………………………………………… (46)
13 Mechanisms ……………………………………………………… (50)
14 Basic Concepts of Mechanisms ………………………………… (53)
15 Introduction to Tribology ……………………………………… (58)
16 Wear and Lubrication ………………………………………… (61)
17 Fundamentals of Mechanical Design ………………………… (66)
18 Machine Designer's Responsibility …………………………… (70)
19 Introduction to Machine Design ……………………………… (74)
20 Some Rules for Mechanical Design …………………………… (78)
21 Material Selection ……………………………………………… (82)
22 Selection of Materials ………………………………………… (86)
23 Lathes …………………………………………………………… (90)
24 Drilling Operations …………………………………………… (95)
25 Machining ……………………………………………………… (101)

26 Machine Tools (107)
27 Metal-Cutting Processes (112)
28 Milling Operations (116)
29 Gear Manufacturing Methods (122)
30 Gear Materials (128)
31 Dimensional Tolerances and Surface Roughness (133)
32 Fundamentals of Manufacturing Accuracy (137)
33 Product Drawings (141)
34 Sectional Views (146)
35 Nontraditional Manufacturing Processes (153)
36 Machining of Engineering Ceramics (157)
37 Mechanical Vibrations (161)
38 Definitions and Terminology of Vibration (165)
39 Residual Stresses (169)
40 Laser-Assisted Machining and Cryogenic Machining Technique (173)
41 Effect of Reliability on Product Salability (177)
42 Reliability Requirements (181)
43 Quality and Inspection (185)
44 Coordinate Measuring Machines (189)
45 Computers in Manufacturing (194)
46 Computer-Aided Design and Manufacturing (198)
47 Computer-Aided Process Planning (201)
48 Process Planning (205)
49 Numerical Control (209)
50 Computer Numerical Control (214)
51 Training Programmers (218)
52 NC Programming (222)
53 Industrial Robots (226)
54 Robotics (231)
55 Basic Components of an Industrial Robot (235)

56	Robotic Sensors	(239)
57	Roles of Engineers in Manufacturing	(244)
58	Manufacturing Technology	(247)
59	Mechanical Engineering in the Information Age	(251)
60	Mechanical Engineering and Mechanical Engineers	(255)
61	How to Write a Scientific Paper	(258)
62	Technical Report Writing	(262)

参考译文 (266)

第一课	力学基本概念	(266)
第三课	力和力矩	(267)
第五课	轴和联轴器	(268)
第七课	连接件和弹簧	(270)
第九课	机械零件的强度	(271)
第十一课	滚动轴承	(273)
第十三课	机构	(274)
第十五课	摩擦学概论	(276)
第十七课	机械设计基础	(277)
第十九课	机械设计概论	(279)
第二十一课	材料选择	(280)
第二十三课	车床	(282)
第二十五课	机械加工	(284)
第二十七课	金属切削加工	(285)
第二十九课	齿轮制造方法	(287)
第三十一课	尺寸公差与表面粗糙度	(291)
第三十三课	产品图样	(292)
第三十五课	特种加工工艺	(294)
第三十六课	工程陶瓷的机械加工	(296)
第三十七课	机械振动	(298)
第三十八课	振动的定义和术语	(299)
第三十九课	残余应力	(301)
第四十一课	可靠性对产品销售的影响	(303)

第四十三课	质量与检测	(305)
第四十五课	计算机在制造业中的应用	(307)
第四十七课	计算机辅助工艺设计	(308)
第四十九课	数字控制	(310)
第五十一课	培训编程人员	(312)
第五十三课	工业机器人	(314)
第五十五课	工业机器人的基本组成部分	(316)
第五十七课	工程师在制造业中的作用	(317)
第五十九课	信息时代的机械工程	(319)
第六十一课	如何撰写科技论文	(321)
Glossary		(324)
主要参考文献		(359)

1 Basic Concepts in Mechanics

The branch of scientific analysis which deals with motions, time, and forces is called mechanics and is made up of two parts, statics and dynamics. Statics deals with the analysis of stationary systems, i.e., those in which time is not a factor, and dynamics deals with systems which change with time.

Forces are transmitted into machine members through mating surfaces, e. g., from a gear to a shaft or from one gear through meshing teeth to another gear (see Fig. 1.1), from a V belt to a pulley, or from a cam to a follower (see Fig. 1.2). It is necessary to know the magnitudes of these forces for a variety of reasons. For example, if the force operating on a journal bearing becomes too high, it will squeeze out the oil film and cause metal-to-metal contact, overheating, and rapid failure of the bearing. If the forces between gear teeth are too large, the oil film may be squeezed out from between them. This could result in flaking and spalling of the metal, noise, rough motion, and eventual failure. In the study of mechanics we are principally interested in determining the magnitude, direction, and location of the forces.

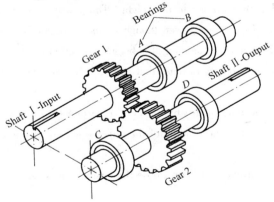

Figure 1.1   Two shafts carrying gears in mesh

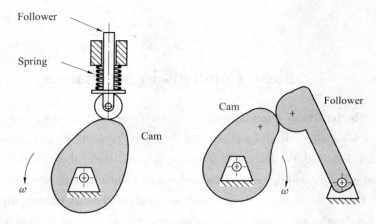

Figure 1.2 Cams and followers

Some of the terms used in mechanics are defined below.

Force Our earliest ideas concerning forces arose because of our desire to push, lift, or pull various objects. So force is the action of one body on another. Our intuitive concept of force includes such ideas as place of application, direction, and magnitude, and these are called the characteristics of a force.

Mass is a measure of the quantity of matter that a body or an object contains. The mass of the body is not dependent on gravity and therefore is different from but proportional to its weight. Thus, a moon rock has a certain constant amount of substance, even though its moon weight is different from its earth weight. This constant amount of substance is called the mass of the rock.

Inertia is the property of mass that causes it to resist any effort to change its motion.

Weight is the force with which a body is attracted to the earth or another celestial body, equal to the product of the object's mass and the acceleration of gravity.

Particle A particle is a body whose dimensions are so small that they may be neglected.

Rigid Body A rigid body does not change size and shape under the action of forces. Actually, all bodies are either elastic or plastic and will be deformed if acted upon by forces. When the deformation of such bodies is small, they are frequently assumed to be rigid, i.e., incapable of deformation, in order to simplify the analysis. A rigid body is an idealization of a real body.

Deformable Body The rigid body assumption cannot be used when internal stresses and strains due to the applied forces are to be analyzed. Thus we consider the body to be capable of deforming. Such analysis is frequently called elastic body analysis, using the additional assumption that the body remains elastic within the range of the applied forces.

Newton's Laws of Motion Newton's three laws are:

Law 1 If all the forces acting on a body are balanced, the body will either remain at rest or will continue to move in a straight line at a uniform velocity.

Law 2 If the forces acting on a body are not balanced, the body will experience an acceleration. The acceleration will be in the direction of the resultant force, and the magnitude of the acceleration will be proportional to the magnitude of the resultant force and inversely proportional to the mass of the body.

Law 3 The forces of action and reaction between interacting bodies are equal in magnitude, opposite in direction, and have the same line of action.

Mechanics deals with two kinds of quantities: scalars and vectors. Scalar quantities are those with which a magnitude alone is associated. Examples of scalar quantities in mechanics are time, volume, density, speed, energy, and mass. Vector quantities, on the other hand, possess direction as well as magnitude. Examples of vectors are displacement, velocity, acceleration, force, moment, and momentum.

Words and Expressions

statics ['stætiks] *n*. 静力学,静止状态
dynamics [dai'næmiks] *n*. 动力学,原动力,动力特性
i.e. 那就是,即(= that is)
mating ['meitiŋ] *n*.;*a*. 配合(的),配套(的),相连(的)

mating surface 配合表面,接触表面

e.g. 例如(= for example)

gear [giə] ***n*** . 齿轮,齿轮传动装置

shaft [ʃɑ:ft] ***n*** . 轴

meshing ['meʃiŋ] ***n*** . 啮合,咬合,钩住

bearing ['bɛəriŋ] ***n*** . 轴承,支承,承载

pulley ['puli] ***n*** . 滑轮,皮带轮

cam [kæm] ***n*** . 凸轮,偏心轮,样板,靠模,仿形板

follower ['fɔləuə] ***n*** . 从动件

spring [spiŋ] ***n*** . 弹簧

magnitude ['mægnitju:d] ***n*** . 大小,尺寸,量度,数值

journal bearing 滑动轴承,向心滑动轴承

squeeze [skwi:z] ***v*** . 挤压,压缩;***n*** . 压榨,挤压

squeeze out 挤压,压出

flaking ['fleikiŋ] ***n*** . 薄片,表面剥落,压碎;***a*** . 易剥落的

spall [spɔ:l] ***v*** . 削,割,打碎,剥落,脱皮;***n*** . 裂片,碎片

intuitive [in'tjuitiv] ***a*** . 直觉的,本能的,天生的

inertia [i'nə:ʃiə] ***n*** . 惯性,惯量,惰性,不活动

celestial [si'lestjəl] ***a*** . 天体的

celestial body 天体

incapable [in'keipəbl] ***a*** . 无能力的,不能的,无用的,无资格的

deformation [ˌdi:fɔː'meiʃən] ***n*** . 变形,形变,扭曲,应变

deformable [di'fɔ:məbl] ***a*** . 可变形的,应变的

acceleration [ækˌseləˈreiʃən] ***n*** . 加速度,加速度值,促进,加快

resultant [ri'zʌltənt] ***a*** . 合的,组合的,总的;***n*** . 合力,合量

scalar ['skeilə] ***n*** . ;***a*** . 数量(的),标量(的)

vector ['vektə] ***n*** . 矢量,向量

displacement [dis'pleismənt] ***n*** . 位移

velocity [vi'lɔsiti] ***n*** . 速度

moment ['məumənt] ***n*** . 力矩

momentum [məu'mentəm] ***n*** . 动量

2 Simple Stress and Strain

In any engineering structure the individual components or members will be subjected to external forces arising from the service conditions or environment in which the component works. If the component or member is in equilibrium, the resultant of the external forces will be zero but, nevertheless, they together place a load on the member which tends to deform that member and which must be reacted by internal forces set up within the material.

There are a number of different ways in which load can be applied to a member. Loads may be classified with respect to time:

(a) A static load is a gradually applied load for which equilibrium is reached in a relatively short time.

(b) A sustained load is a load that is constant over a long period of time, such as the weight of a structure. This type of load is treated in the same manner as a static load; however, for some materials and conditions of temperature and stress, the resistance to failure may be different under short time loading and under sustained loading.

(c) An impact load is a rapidly applied load (an energy load). Vibration normally results from an impact load, and equilibrium is not established until the vibration is eliminated, usually by natural damping forces.

(d) A repeated load is a load that is applied and removed many thousands of times.

(e) A fatigue or alternating load is a load whose magnitude and sign are changed with time.

It has been noted above that external force applied to a body in equilibrium is reacted by internal forces set up within the material. If, therefore, a bar is subjected to a uniform tension or compression, i.e. a force, which is uniformly applied across the cross-section, then the internal forces set up are also distributed uniformly and the bar is said to be subjected to a uniform normal stress, the stress being defined as

$$\text{stress}\,(\sigma) = \frac{\text{load}}{\text{area}} = \frac{P}{A} \qquad (2.1)$$

Stress σ may thus be compressive or tensile depending on the nature of the load and will be measured in units of Newtons per square meter (N/m^2) or multiples of this.

If a bar is subjected to an axial load, and hence a stress, the bar will change in length. If the bar has an original length L and changes in length by an amount δL, the strain produced is defined as follows:

$$\text{strain}\,(\varepsilon) = \frac{\text{change in length}}{\text{original length}} = \frac{\delta L}{L} \qquad (2.2)$$

Strain is thus a measure of the deformation of the material and is non-dimensional, i.e. it has no units; it is simply a ratio of two quantities with the same unit.

Since, in practice, the extensions of materials under load are very small, it is often convenient to measure the strains in the form of strain $\times 10^{-6}$, i.e. microstrain, when the symbol used becomes $\mu\varepsilon$.

Tensile stresses and strains are considered positive in sense. Compressive stresses and strains are considered negative in sense. Thus a negative strain produces a decrease in length.

Figure 2.1 shows a typical stress-strain curve for a material such as ductile low-carbon steel. Ductility refers to the ability of a material to withstand a significant amount stretching before it breaks. Structural steels and some aluminum alloys are examples of ductile metals, and they are often used in machine components because when they become overloaded, they tend to noticeably stretch or bend before they break. On the other end, so-called brittle materials such as cast iron, engineering ceramics, and glass break suddenly and without much prior warning when they are overloaded.

A stress-strain curve for a ductile material is broken down into two regions: the low-strain elastic region (where no permanent deformation remains after a load has been applied and removed) and the high-strain plastic region (where the load is large enough that, upon removal, the material is permanently elongated). For strains below the proportional limit A in Fig. 2.1, stress is proportional to strain. Hooke's applies, it has been shown that

$$E = \frac{stress}{strain} = \frac{\sigma}{\varepsilon} \qquad (2.3)$$

The quantity E is called the elastic modulus, and it has units of force per unit area, such as N/m^2 or GPa. The elastic modulus is a material property, and it is measured simply as the slope of the stress-strain curve in its straight-line region.

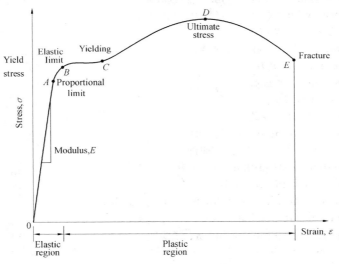

Figure 2.1 Stress-strain curve for a typical low-carbon steel in tension

Elastic modulus E is generally assumed to be the same in tension or compression and for most engineering materials has a high numerical value. Typically, $E = 200 \times 10^9$ N/m^2 for steel, so that it will be observed from Eq. (2.3) that strains are normally very small.

In most common engineering applications strains rarely exceed 0.1%. The actual value of elastic modulus for any material is normally determined by carrying out a standard test on a specimen of the material.

Words and Expressions

stress [stres] *n*. 应力
strain [strein] *n*. 应变
service condition 使用状况,使用条件
component [kəm'pəunənt] *n*. 部件,构件
equilibrium [,i:kwi'libriəm] *n*. 平衡,均衡
internal force 内力
gradually ['grædjuəli] *ad*. 逐渐地
sustained load 长期(施加的)载荷,持续载荷
resistance [ri'zistəns] *n*. 抵抗力,阻力,抗力
impact load 冲击载荷
vibration [vai'breiʃən] *n*. 振动
damping force 阻尼力
repeated load 重复载荷
fatigue load 疲劳载荷
alternating load 交变载荷
uniform ['ju:nifɔ:m] *a*. 均匀的
tension ['tenʃən] *n*. 拉力,拉伸;*v*. 拉伸
compression [kəm'preʃən] *n*. 压力,压缩
multiple ['mʌltipl] *n*. 倍数
non-dimensional ['nɔndi'menʃənəl] *a*. 无量纲的
extension [iks'tenʃən] *n*. 伸长,延长,延期
microstrain [,maikrəu'strein] *n*. 微应变
sense [sens] *n*. 方向
ductile ['dʌktail] *a*. 延性的
ductility [dʌk'tiliti] *n*. 延性
yield stress 屈服应力
yielding ['ji:ldiŋ] *n*. 屈服
proportional limit 比例极限
ultimate stress 极限应力
elesticity [ilæs'tisiti] *n*. 弹性
elastic modulus 弹性模量
permanent ['pə:mənənt] *a*. 永久的,持久的
specimen ['spesimən] *n*. 试件,标本,样品

3 Forces and Moments

When a number of bodies are connected together to form a group or system, the forces of action and reaction between any two of the connecting bodies are called constraint forces. These forces constrain the bodies to behave in a specific manner. Forces external to this system of bodies are called applied forces.

Electric, magnetic, and gravitational forces are examples of forces that may be applied without actual physical contact. A great many of the forces with which we shall be concerned occur through direct physical or mechanical contact.

Force F is a vector. The characteristics of a force are its magnitude, its direction, and its point of application. The direction of a force includes the concept of a line, along which the force is directed, and a sense. Thus, a force is directed positively or negatively along a line of action.

Two equal and opposite forces acting along two noncoincident parallel straight lines in a body cannot be combined to obtain a single resultant force. Any two such forces acting on a body constitute a couple. The arm of the couple is the perpendicular distance between their lines of action, and the plane of the couple is the plane containing the two lines of action.

The moment of a couple is another vector M directed normal to the plane of the couple; the sense of M is in accordance with the right-hand rule for rotation. The magnitude of the moment is the product of the arm of the couple and the magnitude of one of the forces.

A rigid body is in static equilibrium if:

(1) The vector sum of all forces acting upon it is zero.

(2) The sum of the moments of all the forces acting about any single axis is zero.

Mathematically these two statements are expressed as

$$\Sigma F = 0 \quad \Sigma M = 0$$

The term "rigid body" as used here may be an entire machine, several connected parts of a machine, a single part, or a portion of a part. A free-body diagram is a sketch or drawing of the body, isolated from the machine, on which the forces and moments are shown in action (see Fig. 3.1). It is usually desirable to include on the diagram the known magnitudes and directions as well as other pertinent information.

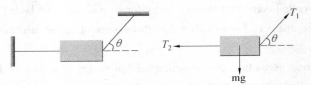

Figure 3.1 An object attached with two strings and its free body diagram

The diagram so obtained is called "free" because the part or portion of the body has been freed from the remaining machine elements and their effects have been replaced by forces and moments. If the free-body diagram is of an entire machine part, the forces shown on it are the external forces (applied forces) and moments exerted by adjoining or connected parts. If the diagram is a portion of a part, the forces and moments acting on the cut portion are the internal forces and moments exerted by the part that has been cut away.

The construction and presentation of clear and neatly drawn free-body diagrams represent the heart of engineering communication. This is true because they represent a part of the thinking process, whether they are actually placed on paper or not, and because the construction of these diagrams is the only way the results of thinking can be communicated to others. You should acquire the habit of drawing free-body diagrams no matter how simple the problem may appear to be. Construction of the diagrams speeds up the problem-solving process and greatly decreases the chances of making mistakes.

The advantages of using free-body diagrams can be summarized as follows:

(1) They make it easy for one to translate words and thoughts and ideas into physical models.

(2) They assist in seeing and understanding all facets of a problem.

(3) They make mathematical relations easier to see or find.

(4) Their use makes it easy to keep track of one's progress and helps in making simplifying assumptions.

(5) The methods used in the solution may be stored for future reference.

(6) They assist your memory and make it easier to explain and present your work to others.

In analyzing the forces in machines we shall almost always need to separate the machine into its individual components and construct free-body diagrams showing the forces that act upon each component. Many of these parts will be connected to each other by kinematic pairs.

Words and Expressions

constraint [kən'streint] *n*. 抑制,限制,制约,约束
constrain [kən'strein] *v*. 强迫,强制,制约,约束,束缚
sense [sens] *n*.;*v*. 显示,方向
noncoincident [ˌnɔnkəu'insidənt] *a*. 不重合的,不一致的,不符合的
parallel ['pærəlel] *a*. 并行的,平行的,相同的;*n*. 平行线
perpendicular [pəːpən'dikjulə] *a*. 垂直的;*n*. 垂直,正交,垂线
product ['prɔdəkt] *n*. 产品,乘积
free-body 自由体,隔离体
free-body diagram 隔离体受力图,隔离体图
sketch [sketʃ] *n*. 草图,简图,示意图,设计图
drawing ['drɔːiŋ] *n*. 绘图,制图,图样
couple ['kʌpl] *n*. 力偶
diagram ['daiəgræm] *n*. 图表,简图,示意图;*v*. 用图表示出
facet ['fæsit] *n*. 面,小平面,事情的某一方面
kinematic [ˌkaini'mætik] *a*. 运动的,运动学的
pair [pɛə] *n*. 一双,一对,一副;*v*. 成对,配合
kinematic pair 运动副

4 Overview of Engineering Mechanics

As we look around us we see a world full of "things": machines, devices, tools; things that we have designed, built, and used; things made of wood, metals, ceramics, and plastics. We know from experience that some things are better than others; they last longer, cost less, are quieter, look better, or are easier to use.

Ideally, however, every such item has been designed according to some set of "functional requirements" as perceived by the designers—that is, it has been designed so as to answer the question, "Exactly what function should it perform?" In the world of engineering, the major function frequently is to support some type of loading due to weight, inertia, pressure, etc. From the beams in our homes to the wings of an airplane, there must be an appropriate melding of materials, dimensions, and fastenings to produce structures that will perform their functions reliably for a reasonable cost over a reasonable lifetime.

In practice, engineering mechanics methods are used in two quite different ways:

(1) The development of any new device requires an interactive, iterative consideration of form, size, materials, loads, durability, safety, and cost.

(2) When a device fails (unexpectedly) it is often necessary to carry out a study to pinpoint the cause of failure and to identify potential corrective measures. Our best designs often evolve through a successive elimination of weak points.

To many engineers, both of the above processes can prove to be absolutely fascinating and enjoyable.

In any "real" problem there is never sufficient good, useful information; we seldom know the actual loads and operating conditions with any precision, and the analyses are seldom exact. While our mathematics may be precise, the overall analysis is generally only approximate, and different skilled people can obtain different solutions. In the study of engineering mechanics, most of the problems will be sufficiently "idealized" to permit unique solutions, but it should be clear that the "real world" is far less idealized, and that you usually will have to perform some idealization in order to obtain a solution.

The technical areas we will consider are frequently called "statics" and "strength of materials," "statics" referring to the study of forces acting on stationary devices, and "strength of materials" referring to the effects of those forces on the structure(deformations, load limits, etc.).

While a great many devices are not, in fact, static, the methods developed here are perfectly applicable to dynamic situations if the extra loadings associated with the dynamics are taken into account. Whenever the dynamic forces are small relative to the static loadings, the system is usually considered to be static.

In engineering mechanics, we will appreciate the various types of approximations that are inherent in any real problem:

Primarily, we will be discussing things which are in "equilibrium," i.e., not accelerating. However, if we look closely enough, everything is accelerating. We will consider many structural members to be "weightless"—but they never are. We will deal with forces that act at a "point"—but all forces act over an area. We will consider some parts to be "rigid"—but all bodies will deform under load.

We will make many assumptions that clearly are false. But these assumptions should always render the problem easier, more tractable. You will discover that the goal is to make as many simplifying assumptions as possible without seriously degrading the result.

Generally there is no clear method to determine how completely, or how precisely, to treat a problem: If our analysis is too simple, we may not get a pertinent answer; if our analysis is too detailed, we may not be able to obtain any answer. It is usually preferable to start with a relatively simple analysis and then add more detail as required to obtain a practical solution.

During the past several decades, there has been a tremendous growth in the availability of computerized methods for solving problems that previously were beyond solution because the time required to solve them would have been prohibitive. At the same time the cost of computers has decreased by orders of magnitude. The computer programs not only remove the drudgery of computation, they allow fairly complicated problems to be solved with ease. Students gain a greater understanding of the subject by simply changing input values and seeing what happens.

Words and Expressions

ceramics [siˈræmiks] ***n***. 陶瓷,陶瓷材料
perceive [pəˈsiːv] ***vt***. 感觉,觉察,发觉,领会,理解,看出
so as to 使得,如此……以至于
meld [meld] ***v***. = merge 融合,汇合,组合,配合,交汇
fastening [ˈfɑːsniŋ] ***n***. 连接,紧固件,连接物
quantitative [ˈkwɔntitətiv] ***a***. 数量的,定量的
interactive [ˌintərˈæktiv] ***a***. 相互作用的,相互影响的,交互的
iterative [ˈitərətiv] ***a***. 反复的,迭代的,重复的
durability [ˌdjuərəˈbiliti] ***n***. 耐久性,持久性,耐用期限
pinpoint [ˈpinpɔint] ***n***. 针尖; ***a***. 极精确的,细致的; ***vt***. 准确定位,正确指出, 确认,强调
evolve [iˈvɔlv] ***v***. 开展,发展,研究出,(经试验研究等)得出
strength of materials 材料力学,也可写作 mechanics of materials
approximation [əprɔksiˈmeiʃən] ***n***. 接近,近似,概算,逼近法
(be) inherent in 为……所固有,是……的固有性质
render [ˈrendə] ***v***. 提出,给予,描绘,表现
tractable [ˈtræktəbl] ***a***. 易处理的,易加工的
degrade [diˈgreid] ***v***. 降低,降级,减低,降解
prohibitive [prəˈhibitiv] ***a***. 不能被人们所接受的,抑制的,昂贵的
order of magnitude 数量级
drudgery [ˈdrʌdʒəri] ***n***. 苦差事,辛苦的工作
with ease 容易地,轻而易举地
greater understanding of the subject 对这个问题(门学科)更深入的理解

5 Shafts and Couplings

Virtually all machines contain shafts. The most common shape for shafts is circular and the cross section can be either solid or hollow (hollow shafts can result in weight savings). Rectangular shafts are sometimes used, as in slotted screwdriver blades (see Fig. 5.1).

—Blade

Figure 5.1 A sloted screwdriver

A shaft must have adequate torsional strength to transmit torque and not be overstressed. It must also be torsionally stiff enough so that one mounted component does not deviate excessively from its original angular position relative to a second component mounted on the same shaft. Generally speaking, the angle of twist should not exceed one degree in a shaft length equal to 20 diameters.

Shafts are mounted in bearings and transmit power through such devices as gears, pulleys, cams and clutches. These devices introduce forces which attempt to bend the shaft; hence, the shaft must be rigid enough to prevent overloading of the supporting bearings. In general, the bending deflection of a shaft should not exceed 0.5 mm per meter of length between bearing supports.

In addition, the shaft must be able to sustain a combination of bending and torsional loads. Thus an equivalent load must be considered which takes into account both torsion and bending. Also, the allowable stress must contain a factor of safety which includes fatigue, since torsional and bending stress reversals occur.

For diameters less than 75 mm, the usual shaft material is cold-rolled steel containing about 0.4 percent carbon. Shafts are either cold-rolled or forged in sizes from 75 mm to 125 mm. For sizes above 125 mm, shafts are forged and machined to the required size. Plastic shafts are widely used for light load applications. One advantage of using plastic is safety in electrical applications, since plastic is a poor conductor of electricity.

Components such as gears and pulleys are mounted on shafts by means of key (see Fig. 32.2). The design of the key and the corresponding keyway in the shaft must be properly evaluated. For example, stress concentrations occur in shafts due to keyways, and the material removed to form the keyway further weakens the shaft.

When a change in diameter occurs in a shaft to create a shoulder against which to locate a machine element (see Fig. 5.2), a stress concentration dependent on the ratio of the two diameters and on the fillet radius is produced. It is recommended that the fillet radius be as large as possible to minimize the stress concentration, but at times the gear, bearing, or other element affects the radius that can be used. For example, the inner ring of a bearing has a factory-produced corner radius (see Fig. 11.1), but it is small. The fillet radius on the shaft must be smaller yet in order for the bearing to be located properly against the shoulder (Fig. 5.2a). When an element with a large chamfer on its bore is located against the shoulder, the fillet radius can be much larger (Fig. 5.2b), and the corresponding stress concentration factor is smaller.

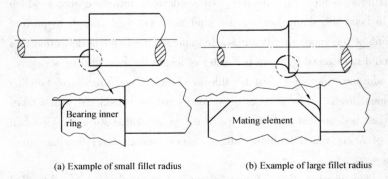

(a) Example of small fillet radius (b) Example of large fillet radius

Figure 5.2 Fillets on shafts

If shafts are run at critical speeds, severe vibrations can occur which can seriously damage a machine. It is important to know the magnitude of these critical speeds so that they can be avoided. As a general rule of thumb, the difference between the operating speed and the critical speed should be at least 20 percent.

Another important aspect of shaft design is the method of directly connecting one shaft to another. This is accomplished by devices such as rigid and flexible couplings.

A coupling is a device for connecting the ends of adjacent shafts. In machine construction, couplings are used to effect a semipermanent connection between adjacent rotating shafts. The connection is permanent in the sense that it is not meant to be broken during the useful life of the machine, but it can be broken and restored in an emergency or when worn parts are replaced.

There are several types of shaft couplings, their characteristics depend on the purpose for which they are used. If an exceptionally long shaft is required in a manufacturing plant or a propeller shaft on a ship, it is made in sections that are coupled together with rigid couplings.

In connecting shafts belonging to separate devices (such as an electric motor and a gearbox), precise aligning of the shafts is difficult and a flexible coupling is used. This coupling connects the shafts in such a way as to minimize the harmful effects of shaft misalignment. Flexible couplings also permit the shafts to deflect under their separate systems of loads and to move freely (float) in the axial direction without interfering with one another. Flexible couplings can also serve to reduce the intensity of shock loads and vibrations transmitted from one shaft to another.

Words and Expressions

coupling ['kʌpliŋ] *n*. 联轴器,连接,耦合
rectangular [rek'tæŋgjulə] *a*. 矩形的,直角的
cross section 截面,横断面,剖面
screw [skru:] *n*. 螺旋丝杆,螺钉
screwdriver 螺丝刀,螺丝起子,改锥

slotted screwdriver blade 一字螺丝刀的扁平头
torsional [ˈtɔːʃənl] *a*. 扭转的,扭力的
torque [ˈtɔːk] *n*. 转矩,扭矩,扭转
mounted [ˈmauntid] *a*. 安装好的,固定好的,安装在……上的
deviate [ˈdiːvieit] *v*. 偏离,背离,偏差,偏移
deviate from 偏离,与……有偏差
twist [twist] *v*. 使扭转,扭,使转动
clutch [klʌtʃ] *v*.; *n*. 抓住,捏紧,离合器
bending [ˈbendiŋ] *n*. 弯曲,弯曲度,挠曲,挠曲度
deflection [diˈflekʃən] *n*. 偏转,偏差,挠曲,弯曲,挠度
reversal [riˈvəːsəl] *n*. 颠倒,相反,反向,改变方向,倒转
cold-roll *n*.; *v*. 冷轧,冷轧机
forge [fɔːdʒ] *n*.; *v*. 锻造,打制,锻工车间
key [kiː] *n*. 键
keyway 键槽
fillet [ˈfilit] *n*. 圆角,过渡圆角
fillet radius 内圆角半径,齿根圆角半径,过渡圆半径
stress concentration 应力集中
corner radius 圆角半径,外圆角半径
chamfer [ˈtʃæmfə] *n*. 斜面,倒角,倒棱
bearing inner ring 轴承内圈
mating element 配合零件,互相配合的零件
critical speed 临界转速,临界速度
adjacent [əˈdʒeisənt] *a*. 邻近的,接近的,相邻的
semipermanent [ˌsemiˈpəːmənent] *a*. 半永久性的
propeller [prəˈpelə] *n*. 螺旋桨,推进器
gearbox [ˈgiəbɔks] *n*. 齿轮箱,变速箱
flexible coupling 弹性联轴器,挠性联轴器
in such a way as 以这样一种方式
shock [ʃɔk] *n*. 冲击,冲撞,打击

6 Shaft Design

A shaft is a rotating or stationary member, usually of circular cross section, having mounted upon it such elements as gears, pulleys, flywheels, and other power-transmission elements.

If a shaft carries several gears or pulleys, different sections of the shaft will be subjected to different torques, because the total power delivered to the the shaft is taken off piecemeal at various points. Figure 6.1 shows a shaft with an input at one location (point A) and several outputs (points B, C, and D) on either side of the input. Gears or pulleys can be used to supply power to the shaft or to take power off the shaft. Hence one must note the amount of torque on each part of the shaft. Then study the distribution of the bending moment, preferably sketching (freehand is all right) the shear force and bending moment diagrams.

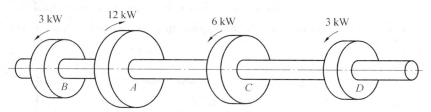

Figure 6.1 Power take offs

From this preliminary examination, which is a problem in mechanics, we note the section where the bending moment is a maximum and the section where the torque is a maximum. If these maximums occur at the same section, the diameter needed for that section is determined—and used for the entire shaft when the diameter is to be constant. If the maximums do not occur at the same section, determine the diameter for the section of maximum torque and also for the section of maximum bending moment, and use the larger value.

The diameter of shaft is often varied from point to point, sometimes for structural reasons (see Fig 6.2). In this case, check the stress or determine the size needed for each section. The designer makes certain that all sections of the shaft are subjected to safe stresses, taking due note of fillets, holes, keyways, and other stress raisers.

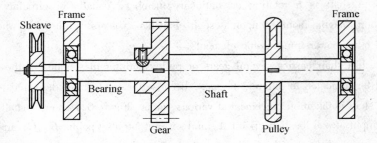

Figure 6.2 A stepped shaft with various elements attached

Hollow round shafts (see Fig. 6.3) sometimes serve a useful purpose, usually in large sizes, though they are more expensive than solid ones. They have the advantages of being stronger and stiffer, weight for weight, because the outer fibers are more effective in resisting the applied moments, and they respond better to heat treatment because quenching can proceed outward as well as inward.

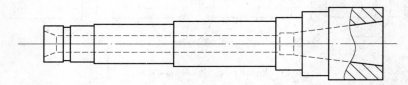

Figure 6.3 A hollow shaft

Deflection is a significant consideration in the design of shafts. Criteria for the limiting torsional deflection vary from 0.25° per meter of length for machinery shafts to 1° per meter or 1° in a length of 20 diameters for transmission shafting. Even short shafts become special problems in rigidity when the load is applied in impulses, as on an automobile crankshaft

(Fig. 6.4). The impulses produce a torsional vibration, usually compensated by torsional-vibration dampers in an automotive engine.

Figure 6.4　A crankshaft

Data on permissible values of deflections are rare, probably because the range of values would be large and each situation has its own peculiarities. An old rule of thumb for transmission shafting is that the deflection should not exceed $0.0005L$, where L is the shaft length between supports; although greater stiffness may be desired. Preferably, on transmission shafts, the pulleys and gears should be located close to bearings in order to minimize moments. Excessive lateral shaft deflection can hamper gear performance and cause objectionable noise. The associated angular deflection can be very destructive to non-self-aligning bearings. Self-aligning bearings (Fig. 6.5) may eliminate this trouble if the deflection is otherwise acceptable.

(a) Self-aligning roller bearing　　　(b) Self-aligning ball bearing

Figure 6.5　Self-aligning bearings

On machine tools (lathes, milling machines, etc.), rigidity is a special concern because of its relation to accuracy. If a shaft supports a gear, deflection is more of a consideration than if it carries a V-belt pulley.

The center of mass of a symmetric, rotating shaft does not coincide with its center of rotation. All shafts, even in the absence of external load, will deflect during rotation. The unbalanced mass of the rotating shaft causes deflection that will create resonant vibration at certain speeds, known as the critical speeds. A shaft is said to reach critical speed when the speed of its rotation corresponds to one of its natural frequencies.

Some shafts are supported by three or more bearings, which means that the problem is statically indeterminate. Textbooks on strength of materials give methods of solving such problems. The design effort should be in keeping with the economics of a given situation. For example, if one line shaft supported by three or more bearings is needed, it probably would be cheaper to make conservative assumptions as to moments and design it as though it were statically determinate. The extra cost of an oversize shaft may be less than the extra cost of an elaborate design analysis.

Words and Expressions

piecemeal ['piːsmiːl] *ad.*; *a.* 逐点,逐渐,逐段,一部分一部分地
take off 取走,输出,移动,除去
power take off 动力输出
shear [ʃiə] *v.* 剪切,切断; *n.* 剪切,剪力
shear force diagram 剪力图
bending moment diagram 弯矩图
freehand ['friːhænd] *a.* 徒手画的
take due note of 认真地注意到
sheave [ʃiːv] *n.* 带轮,有槽的带轮(grooved pulley), V 带轮(V-belt pulley)
stepped shaft 阶梯轴
weight for weight 就重量比来说,单就重量来说
quenching ['kwentʃiŋ] *n.* 淬火
shafting ['ʃɑːftiŋ] *n.* 轴系,轴
impulse ['impʌls] *n.* 冲击,碰撞,脉冲,脉动

crankshaft [ˈkræŋkʃɑːft] *n*. 曲轴
damper [ˈdæmpə] *n*. 阻尼器,减振器,缓冲器
torsional-vibration damper 扭振阻尼器
permissible [pəˈmisəbl] *a*. 容许的,许可的,安全的
peculiarity [piˌkjuːliˈæriti] *n*. 独特性,特色,特殊的东西
journal [ˈdʒəːnl] *n*. 轴颈,辊颈,枢轴
lubrication [ˌljuːbriˈkeiʃən] *n*. 润滑,润滑作用
lateral [ˈlætərəl] *a*. 横向的
hamper [ˈhæmpə] *v*. 妨碍,阻碍
objectionable noise 令人讨厌的噪音
destructive [disˈtrʌktiv] *a*. 有害的,造成损害的,毁坏的
self-aligning bearing 调心轴承,自位轴承,自动定心轴承
self-aligning roller bearing 调心滚子轴承
otherwise acceptable 除此之外是可以被接受的
machine tool 机床
lathe [leið] *n*.; *v*. 车床,用车床加工,车削
milling [ˈmiliŋ] *n*. 铣削
milling machine 铣床
symmetric [siˈmetrik] *a*. 对称的,平衡的; *n*. 对称
coincide [ˌkəuinˈsaid] *v*. (点,面积,轮廓等)与……重合,与……一致,相同
center of mass 质心
unbalanced mass 不均衡质量
resonant [ˈrezənənt] **vibration** 共振
natural frequeney 固有频率,又称"自振频率"
statically indeterminate 超静定的,静不定的
design effort 设计工作
given situation 特定情况,特定情况下
line shaft 主传动轴,动力轴
make conservative assumptions as to moments 对力矩作出保守的假设

7 Fasteners and Springs

Fasteners are devices which permit one part to be joined to a second part and, hence, they are involved in almost all designs. The acceptability of any product depends not only on the selected components, but also on the means by which they are fastened together. The principal purposes of fasteners are to provide the following design features:

(1) Disassembly for inspection and repair;

(2) Modular design where a product consists of a number of subassemblies. Modular design aids manufacturing as well as transportation.

There are three main classifications of fasteners, which are described as follows:

(1) Removable. This type permits the parts to be readily disconnected without damaging the fastener. An example is the ordinary nut-and-bolt fastener.

(2) Semipermanent. For this type, the parts can be disconnected, but some damage usually occurs to the fastener. One such example is a cotter pin (Fig. 7.1).

Figure 7.1 Cotter pin

(3) Permanent. When this type of fastener is used, it is intended that the parts will never be disassembled. Examples are riveted and welded joints.

The following factors should be taken into account when selecting fasteners for a given application:

(1) Primary function;

(2) Appearance;

(3) A large number of small size fasteners versus a small number of large size fasteners (an example is bolts);

(4) Operating conditions such as vibration, loads and temperature;
(5) Frequency of disassembly;
(6) Adjustability in the location of parts;
(7) Types of materials to be joined;
(8) Consequences of failure or loosening of the fastener.

The importance of fasteners can be realized when referring to any complex product. In the case of the automobile, there are literally thousands of parts which are fastened together to produce the total product. The failure or loosening of a single fastener could result in a simple nuisance such as a door rattle or in a serious situation such as a wheel coming off. Such possibilities must be taken into account in the selection of the type of fastener for the specific application.

Springs are mechanical members which are designed to give a relatively large amount of elastic deflection under the action of an externally applied load. Hooke's Law, which states that deflection is proportional to load, is the basis of behavior of springs. However, some springs are designed to produce a nonlinear relationship between load and deflection. The following is a list of the important purposes and applications of springs:

(1) Control of motion in machines. This category represents the majority of spring applications such as operating forces in clutches and brakes. Also, springs are used to maintain contact between two members such as a cam and its follower.

(2) Reduction of transmitted forces as a result of impact or shock loading. Applications here include automotive suspension system springs and bumper springs.

(3) Storage of energy. Applications in this category are found in clocks, movie cameras and lawn mowers.

(4) Measurement of force. Scales used to weigh people is a very common application for this category.

The three major classifications of springs are compression, extension, and torsion (see Fig. 7.2). Compression and extension springs are the springs most often used. Deflection is linear in these types. Torsion springs are characterized by angular instead of linear deflection. Leaf springs are of the simple beam or cantilever type, their deflection is linear.

Most springs are made of steel, although silicon bronze, brass and

(a) Compression spring   (b) Extension spring   (c) Torsion spring

Figure 7.2   Helical springs

beryllium copper are also used. Springs are universally made by companies which specialize in the manufacture of springs. The helical spring is the most popular type of spring; torsion bar springs (Fig. 7.3) and leaf springs are also widely used. If the wire diameter (assuming a helical spring) is less than 8 mm, the spring will normally be cold-wound from hard-drawn or oil-tempered wire. For larger diameters, springs are formed using hot-rolled bar.

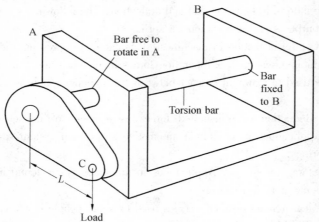

Figure 7.3   Torsion bar spring

It is good practice to consult with a spring manufacturing company when selecting a spring design, especially if high loads or temperatures are to be encountered, or if stress reversals occur or corrosion resistance is required. To properly select a spring, a complete study of the spring requirements, including space limitations, must be undertaken. Many different types of special springs are available to satisfy unusual requirements or applications.

Words and Expressions

fastener ['fɑːsnə] *n*. 紧固件,连接件
acceptability [əkˌseptə'biliti] *n*. 可接受性,接受,合格,满意
disassembly [ˌdisə'sembli] *n*. 拆卸,拆除,拆开,分解,解体
modular ['mɔdjulə] *a*. 模数的,制成标准组件的,预制的,组合的
subassembly ['sʌbə'sembli] *n*. 组件,部件,局部装配,组件装配
nut [nʌt] *n*. 螺帽,螺母; *v*. 拧螺母,装螺母
appearance [ə'piərəns] *n*. 出现,外观,外貌,外部特征
versus ['vəːsəs] *prep*. 与……比较,……与……的关系曲线,作为……的函数
cotter ['kɔtə] *n*. 栓,开口销,楔; *v*. 用销固定
cotter pin 扁销,开口销
rivet ['rivit] *n*. 铆钉; *v*. 铆接,固定,钉铆钉
literally ['litərəli] *ad*. 字面上地,逐字地,实在是,真正地,完全地
nuisance ['njuːsns] *n*. 麻烦事情,讨厌的事,障碍,刺激,损害
rattle ['rætl] *v*. 发出喀啦声,发硬物震动声; *n*. 喀啦声
spring [spriŋ] *n*. 弹簧,发条
proportional [prə'pɔːʃnl] *a*. (成正)比例的,平衡的,相称的
suspension [səs'penʃən] *n*. 悬挂,悬置,悬挂装置
bumper ['bʌmpə] *n*. 保险杠,缓冲器,阻尼器,减震器
mower ['məuə] *n*. 割草机
silicon ['silikən] *n*. 硅
torsion bar spring 扭杆弹簧(即一端固定而另一端与工作部件连接的杆形弹簧,主要作用是靠扭转弹力来吸收振动能量)
leaf spring 板弹簧
brass [brɑːs] *n*. 黄铜
bronze [brɔnz] *n*. 青铜
beryllium [bə'riljəm] *n*. 铍
helical ['helikəl] *n*.; *a*. 螺线,螺旋线,螺旋状
torsion ['tɔːʃən] *n*. 扭转
temper ['tempə] *v*.; *n*. 回火,一种热处理方式
corrosion [kə'rəuʒən] *n*. 腐蚀,侵蚀,锈蚀

8 Belt Drives and Chain Drives

In addition to gears, belts and chains are in widespread use. Belts and chains represent the major types of flexible power transmission elements. Figure 8.1 shows a typical industrial application of these elements combined with a gear-type speed reducer. This application illustrates where belts, gear drives, and chains are each used to best advantage.

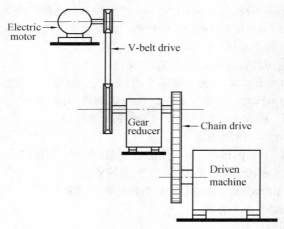

Figure 8.1 Sketch of combination drive

Rotary power is usually developed by the electric motor, but motors typically operate at too high a speed and deliver too low a torque to be appropriate for the final drive application. For a given power transmission, the torque is increased in proportion to the amount that rotational speed is reduced. So some speed reduction is often desirable. In general, belt drives are applied where the rotational speeds are relatively high, as on the first stage of speed reduction from an electric motor or engine. A smaller driving pulley is attached to the motor shaft, while a larger diameter pulley is attached to a parallel shaft that operates at a correspondingly lower speed. Pulleys for belt drives are also called sheaves. The linear speed of a belt is usually 10 ~ 30 m/s, which results in relatively low tensile forces in the belt.

There are four main belt types: flat, round, V, and synchronous. Flat and round belts may be used for long center distances between the pulleys in a belt drive. On the other hand, V and synchronous belts are employed for limited shorter center distance. Excluding synchronous belts, there is some slip between the belt and the pulley, which is usually made of cast iron or steel.

A widely used type of belt, particularly in industrial drives and vehicular applications, is the V-belt drive (Fig. 8.2). Another type of V-belt drive is the poly V-belt drive (Fig. 8.3). The V-shape causes the belt to wedge tightly into the groove of the sheave, increasing friction and allowing high torques to be transmitted before slipping occurs.

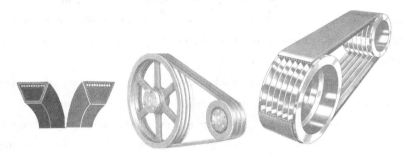

Figure 8.2　V-belts and multiple V-belt drive　　Figure 8.3　Poly V-belt drive

Flat belt drives produce very little noise and absorb more vibration from the system than either V-belt or other drives. A flat belt drive has an efficiency of around 98%, which is nearly the same as for a gear drive. A V-belt drive can transmit more power than a flat belt drive. However, the efficiency of a V-belt drive varies between 70% and 96%.

A synchronous belt, sometimes called timing belt, has evenly spaced teeth on the inside circumference (see Fig. 8.4). A synchronous belt does not slip and hence transmits power at a constant angular velocity ratio.

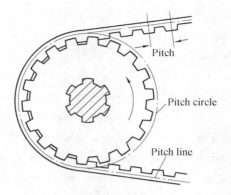

Figure 8.4 Synchronous belt drive

However, if very large ratios of speed reduction are required in the drive, gear reducers are desirable because they can typically accomplish large reductions in a rather small package. The output shaft of the gear-type speed reducer is generally at low speed and high torque. If both speed and torque are satisfactory for the application, it could be directly coupled to the driven machine.

Since gear reducers are available only at discrete reduction ratios, the output must often be reduced more before meeting the requirements of the machine. At the low-speed, high-torque condition, chain drives become desirable. The high torque causes high tensile forces to be developed in the chain. The elements of the chain are typically metal, and they are sized to withstand the high forces. When transmitting power between rotating shafts, the links of chains are engaged in toothed wheels called sprockets to provide positive (no slip) drive. Figure 8.5 shows a typical chain drive.

Chain drives combine some of the more advantageous features of belt and gear drives. Chain drives provide almost any speed ratio. Their chief advantage over gear drives is that chain drives can be used with arbitrary center distances. Compared with belt drives, chain drives offer the advantage of positive drive and therefore greater power capacity.

The most common type of chain is the roller chain, in which the roller on

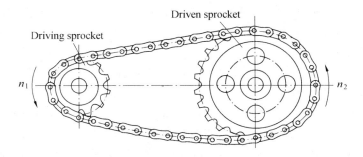

Figure 8.5 Roller chain drive

each bushing (see Fig. 8.6) provides exceptionally low friction between the chain and the sprockets. Of its diverse applications, the most familiar is the roller chain drive on a bicycle. A roller chain is generally made of hardened steel and sprockets are generally made of steel or cast iron. Nevertheless, stainless steel and bronze chains are obtainable where corrosion resistance is needed.

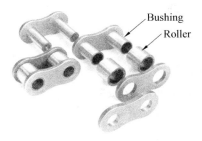

Figure 8.6 Portion of a roller chain

Words and Expressions

belt drive 带传动(由柔性带和带轮组成传递运动和(或)动力的机械传动)
chain drive 链传动(利用链与链轮轮齿的啮合来传递动力和运动的机械传动)
flexible element 挠性件,挠性元件
flexible transmission 挠性传动
power transmission 动力传动
speed reducer 减速器

to best advantage 以最好的方式,效果最佳的
driven machine 从动机,被驱动的机器
first stage 第一级
driving pulley 主动带轮(传动中用于驱动带运动的带轮)
linear speed 线速度
flat belt 平带(横截面为矩形或近似为矩形的传动带,其工作面为宽平面)
round belt 圆带(横截面为圆形或近似为圆形的传动带)
V-belt V带(横截面为等腰梯形或近似为等腰梯形的传动带,其工作面为两侧面)
synchronous ['siŋkrənəs] *a*. 同时的,同步的
synchronous belt 同步带(内表面或者内、外表面具有等距横向齿的环形传动带)
multiple V-belt drive 多根V带传动
poly V-belt 多楔带(以平带为基体、内表面具有等距纵向楔的环形传动带)
timing belt 同步齿型带
circumference [sə'kʌmfərəns] *n*. 圆周,周围
inside circumference 内周边
pitch [pitʃ] *n*. 节距,螺距
pitch circle 节圆
pitch line 节线
angular velocity 角速度
discrete [dis'kri:t] *a*. 离散的,不连续的
reduction ratio 减速比
sprocket ['sprɔkit] *n*. 链轮
positive drive 强制传动,正传动
driving sprocket 主动链轮
driven sprocket 从动链轮
power capacity 功率,额定功率
roller chain 滚子链(组成零件中具有回转滚子,且滚子表面在啮合时直接与链轮齿接触的链条)
bushing ['buʃiŋ] *n*. 套筒
hardened steel 淬硬钢,淬火钢

9 Strength of Mechanical Elements

One of the primary considerations in designing any machine or structure is that the strength must be sufficiently greater than the stress to assure both safety and reliability. To assure that mechanical parts do not fail in service, it is necessary to learn why they sometimes do fail. Then we shall be able to relate the stresses with the strengths to achieve safety.

Ideally, in designing any machine element, the engineer should have at his disposal the results of a great many strength tests of the particular material chosen. These tests should have been made on specimens having the same heat treatment, surface roughness, and size as the element he proposes to design; and the tests should be made under exactly the same loading conditions as the part will experience in service. This means that, if the part is to experience a bending load, it should be tested with a bending load. If it is to be subjected to combined bending and torsion, it should be tested under combined bending and torsion. Such tests will provide very useful and precise information. They tell the engineer what factor of safety to use and what the reliability is for a given service life. Whenever such data are available for design purposes, the engineer can be assured that he is doing the best possible job of engineering.

The cost of gathering such extensive data prior to design is justified if failure of the part may endanger human life, or if the part is manufactured in sufficiently large quantities. Automobiles and refrigerators, for example, have very good reliabilities because the parts are made in such large quantities that they can be thoroughly tested in advance of manufacture. The cost of making these tests is very low when it is divided by the total number of parts manufactured.

You can now appreciate the following four design categories:

(1) Failure of the part would endanger human life, or the part is made in

extremely large quantities; consequently, an elaborate testing program is justified during design.

(2) The part is made in large enough quantities so that a moderate series of tests is feasible.

(3) The part is made in such small quantities that testing is not justified at all; or the design must be completed so rapidly that there is not enough time for testing.

(4) The part has already been designed, manufactured, and tested and found to be unsatisfactory. Analysis is required to understand why the part is unsatisfactory and what to do to improve it.

It is with the last three categories that we shall be mostly concerned. This means that the designer will usually have only published values of yield strength, ultimate strength, and percentage elongation. With this meager information the engineer is expected to design against static and dynamic loads, biaxial and triaxial stress states, high and low temperatures, and large and small parts! The data usually available for design have been obtained from the simple tension test, where the load was applied gradually and the strain given time to develop. Yet these same data must be used in designing parts with complicated dynamic loads applied thousands of times per minute. No wonder machine parts sometimes fail.

To sum up, the fundamental problem of the designer is to use the simple tension-test data and relate them to the strength of the part, regardless of the stress state or the loading situation.

It is possible for two metals to have exactly the same strength and hardness, yet one of these metals may have a superior ability to absorb overloads, because of the property called ductility. Ductility is measured by the percentage elongation which occurs in the material at fracture. The usual dividing line between ductility and brittleness is 5 percent elongation. A material having less than 5 percent elongation at fracture is said to be brittle, while one having more is said to be ductile.

The elongation of a material is usually measured over 50 mm gage length (see Fig. 9.1). Since this is not a measure of the actual strain, another method of determining ductility is sometimes used. After the specimen has been fractured, measurements are made of the area of the cross section at the fracture. Ductility can then be expressed as the percentage reduction in cross-sectional area.

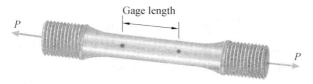

Figure 9.1 Gage length for tensile specimen

The characteristic of a ductile material which permits it to absorb large overloads is an additional safety factor in design. Ductility is also important because it is a measure of that property of a material which permits it to be cold-worked. Such operations as bending and drawing are metal-processing operations which require ductile materials.

When a material is to be selected to resist wear, erosion, or plastic deformation, hardness is generally the most important property. Several methods of hardness testing are available, depending upon which particular property is most desired. The four hardness numbers in greatest use are the Brinell, Rockwell, Vickers, and Knoop.

Most hardness-testing systems employ a standard load which is applied to a ball or pyramid in contact with the material to be tested. The hardness is then expressed as a function of the size of the resulting indentation, as shown in Table 9.1. This means that hardness is an easy property to measure, because the test is nondestructive and test specimens are not required. Usually the test can be conducted directly on an actual machine element.

Words and Expressions

reliability [rilaiə'biliti] *n*. 可靠性, 安全性
sufficiently [sə'fiʃəntli] *ad*. 充足地

at one's disposal 供……使用,归……支配,摆在……面前
surface roughness 表面粗糙度
specimen [ˈspesimin] ***n***. 样品,样本,试样,试件
justify [ˈdʒʌstifai] ***v***. 证明……是正当的,认为……有理由
feasible [ˈfiːzəbl] ***a***. 可行的,做得到的,行得通的,可能的
category [ˈkætigəri] ***n***. 种类,类别,类型,等级
elongation [ˌiːlɔŋˈgeiʃən] ***n***. 拉伸,伸长,延长,延伸率
biaxial [baiˈæksiəl] ***a***. 二轴的,二维的
triaxial [traiˈæksiəl] ***a***. 三维的,三轴的,空间的
meager [ˈmiːgə] ***a***. 贫乏的,不足的,不充分的,量少的
ductility [dʌkˈtiliti] ***n***. 延展性,可锻性,韧性,可塑性
ductile [ˈdʌktail] ***a***. 可延展的,可锻的,可塑的,易变形的
brittle [ˈbritl] ***a***. 脆性的,易碎的,易损坏的
cross-sectional [ˈkrɔsˈsekʃənəl] ***a***. 横截面的,剖面的
gage length 标距(长度),计量长度,也可写为 gauge length
cold-work [ˈkəuldˈwəːk] ***v***. 冷变形加工
erosion [iˈrəuʒən] ***n***. 腐蚀,蚀除,磨损,磨蚀
pyramid [ˈpirəmid] ***n***. 棱锥,四面体; ***v***. 成角锥形
indentation [ˌindenˈteiʃən] ***n***. 压痕,凹痕
nondestructive [ˈnɔndisˈtrʌktiv] ***a***. 非破坏性的,无损的
indenter [inˈdentə] ***n***. (硬度试验)压头
top view 俯视图(由上向下投射所得的视图)
side view 侧视图,侧面图
formula [ˈfɔːmjulə] ***n***. 公式,方案,规则
hardness number 硬度值
tungsten carbide 碳化钨,硬质合金
cone [kəun] ***n***. 圆锥体
Brinell [briˈnel] **hardness** 布氏硬度
Vickers [ˈvikəs] **hardness** 维氏硬度
Knoop [nuːp] **hardness** 努氏硬度
Rockwell [ˈrɔkwel] **hardness** 洛氏硬度

Table 9.1 Comparison of hardness tests

Test	Indenter	Shape of indentation (Side view)	Shape of indentation (Top view)	Load	Formula for hardness number
Brinell	10-mm sphere of steel or tungsten carbide			P	$\mathrm{BHN} = \dfrac{2P}{\pi D[D - \sqrt{D^2 - d^2}]}$
Vickers	Diamond pyramid	136°		P	$\mathrm{VHN} = 1.72 P / d_1^2$
Knoop microhardness	Diamond pyramid	$l/b = 7.11$ $b/t = 4.00$		P	$\mathrm{KHN} = 14 P / l^2$
Rockwell A C D	Diamond cone	120°		60kg 150kg 100kg	$R_A =$ $R_C =$ $R_D =$ $\bigg\} 100 - 500 t$

10 Physical Properties of Materials

One of the important considerations in material selection is their physical properties (that is, density, melting point, specific heat, thermal conductivity, thermal expansion, and corrosion resistance). Physical properties can have several important influences on manufacturing and the service life of components. For example, high-speed machine tools require lightweight components to reduce inertial forces and, thus, keep machines from excessive vibration.

1. Density The density of a material is its mass per unit volume. Another term is specific gravity, which expresses a material's density in relation to that of water, and thus, it has no units. Weight saving is important particularly for aircraft and aerospace structures, for automotive bodies and components, and for other products where energy consumption and power limitations are major concerns. Substitution of materials for the sake of weight saving and economy is a major factor in the design both of advanced equipment and machinery and of consumer products, such as automobiles.

2. Melting Point The melting point of a metal depends on the energy required to separate its atoms. The melting temperature of a metal alloy can have a wide range (depending on its composition) and is unlike that of a pure metal, which has a definite melting point. The temperature range within which a component or structure is designed to function is an important consideration in the selection of materials.

The melting point of a metal has a number of indirect effects on manufacturing operations. Because the recrystallization temperature of a metal is related to its melting point, operations such as annealing and heat treating and hot working require a knowledge of the melting points of the materials involved.

3. **Specific Heat**   A material's specific heat is the energy required to raise the temperature of a unit mass by one degree. Alloying element have a relatively minor effect on the specific heat of metals. The temperature rise in a workpiece resulting from machining operations is a function of the work done and of the specific heat of the workpiece material. Temperature rise in a workpiece, if excessive, can decrease product quality by adversely affecting its surface roughness and dimensional accuracy, can cause excessive tool wear, and can result in undesirable metallurgical changes in the material.

4. **Thermal Conductivity**   Thermal conductivity indicates the rate at which heat flows within and through a material. Metals generally have high thermal conductivity, while ceramics and plastics have poor conductivity.

When heat is generated by plastic deformation or due to friction, the heat should be conducted away at a rate high enough to prevent a severe rise in temperature. The main difficulty experienced in machining titanium, for example, is caused by its very low thermal conductivity. Low thermal conductivity can also result in high thermal gradients and, in this way, cause inhomogeneous deformation of workpieces in metalworking processes.

5. **Thermal Expansion**   The thermal expansion of materials can have several significant effects, particularly the relative expansion or contraction of different materials in assemblies such as moving parts in machinery that require certain clearances for proper functioning.

Shrink fits utilize thermal expansion and contraction. For example, a part, such as a flange (see Fig. 10.1), is to be installed over a shaft. It first is heated and then slipped over a shaft which is at room temperature. When allowed to cool, the part shrinks and the assembly effectively becomes an integral component.

6. **Corrosion Resistance**   Metals, ceramics, and plastics all are subject to forms of corrosion. The word corrosion itself usually refers to the deterioration of metals and ceramics, while similar phenomena in plastics generally are called degradation. Corrosion leads not only to surface deterioration of components and structures but also reduces their strength and

structural integrity.

Corrosion resistance is an important aspect of material selection for applications in the chemical, food, and petroleum industries, as well as in manufacturing operations. In addition to various possible chemical reactions from the elements and compounds present, environmental corrosion of components and structures is a major concern, particularly at elevated temperatures and other transportation vehicles.

Figure 10.1　Flanges

Resistance to corrosion depends on the composition of the material and on the particular environment. Corrosive media may be chemicals (acids, alkalis, and salts), the environment (oxygen, moisture, pollution, and acid rain), and water (fresh or salt water). Nonferrous metals, stainless steels, and nonmetallic materials generally have high corrosion resistance. Steels and cast irons usually have poor resistance and must be protected by various coatings and surface treatments.

The usefulness of some level of oxidation is exhibited in the corrosion resistance of aluminum, titanium, and stainless steel. Aluminum develops a thin (a few atomic layers), strong, and adherent hard-oxide film (Al_2O_3) that better protects the surface from further environmental corrosion. Titanium develops a film of titanium oxide (TiO_2). A similar phenomenon occurs in stainless steels, which (because of the chromium present in the alloy) develop a protective film on their surfaces. When the protective film is scratched and exposes the metal underneath, a new oxide film begins to form.

Words and Expressions

physical property 物理性质，物理性能
density ['densiti] *n*. 密度
melting point 熔点
specific heat 比热
thermal conductivity 热导率，导热率，导热系数
thermal expansion 热膨胀
flange [flændʒ] *n*. 法兰，法兰盘
corrosion resistance 耐蚀性
lightweight ['laitweit] *a*. 轻的，重量轻的
specific gravity 比重
recrystallization [ri,kristəlai'zeiʃən] *n*. 再结晶
annealing [æ'ni:liŋ] *v*. 退火
hot working 热加工
metallurgical [,metə'lə:dʒikəl] *a*. 冶金学的
thermal gradient 热梯度，温度梯度
inhomogeneous [,inhɔmə'dʒi:niəs] *a*. 不均匀的
metalworking ['metəlwə:kiŋ] *n*. 金属加工
shrink fit 冷缩配合，热装配合
deterioration [di,tiəriə'reiʃən] *n*. 恶化，变质，退化
surface deterioration 表面劣化
degradation [,degrə'deiʃən] *n*. 降解，老化，退化
fresh water 清水，淡水
nonferrous metal 有色金属，非铁金属
nonmetallic ['nɔnmi'tælik] *a*. 非金属的 *n*. 非金属物质
titanium [tai'teinjəm] *n*. 钛
stainless steel 不锈钢
chromium ['krəumjəm] *n*. 铬
protective film 保护膜
adherent [əd'hiərənt] *a*. 粘附的，附着的

11 Rolling Bearings

Rolling bearings can carry radial, thrust or combination of the two loads. Accordingly, most rolling bearings are categorized in one of the three groups: radial bearings for carrying loads that are primarily radial, thrust bearings for supporting loads that are primarily axial, and angular contact bearings or tapered roller bearings for carrying combined radial and axial loads. Figure 11.1 shows a common single-row, deep groove ball bearing. The bearing consists of an inner ring, an outer ring, the balls and the separator (also known as retainer). To increase the contact area and hence permit larger loads to be carried, the balls run in curvilinear grooves in the rings called raceway. The radius of the raceway is very little larger than the radius of the ball. This type of bearing can stand a radial load as well as some thrust load. Some other types of rolling bearings are shown in Fig. 11.2.

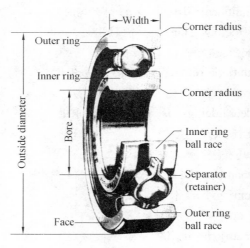

Figure 11.1 Deep groove ball bearing

Figure 11.2 Some types of rolling bearings: (a) Thrust ball bearing;
(b) Tapered roller thrust bearing; (c) Linear bearing.

    The concern of a machine designer with ball and roller bearings is fivefold as follows: (a) life in relation to load; (b) stiffness, i.e. deflections under load; (c) friction; (d) wear; (e) noise. For moderate loads and speeds the correct selection of a standard bearing on the basis of load rating will usually secure satisfactory performance. The deflection of the bearing elements will become important where loads are high, although this is usually of less magnitude than that of the shafts or other components associated with the bearing. Where speeds are high special cooling arrangements become necessary which may increase frictional drag. Wear is primarily associated with the introduction of contaminants, and sealing arrangements must be chosen with regard to the hostility of the environment.

    Because the high quality and low price of ball and roller bearings depends on quantity production, the task of the machine designer becomes one of selection rather than design. Rolling bearings are generally made with steel which is through-hardened to about 900 HV. Owing to the high stresses involved, a predominant form of failure should be metal fatigue, and a good deal of work is currently in progress intended to improve the reliability of this type of bearing. Design can be based on accepted values of life and it is general practice in the bearing industry to define the load capacity of the bearing as that value below which 90 per cent of a batch will exceed a life of one million revolutions.

    Notwithstanding the fact that responsibility for the basic design of ball and roller bearings rests with the bearing manufacturer, the machine designer must form a correct appreciation of the duty to be performed by the bearing and be concerned not only with bearing selection but with the conditions for correct installation.

The fit of the bearing rings onto the shafts or onto the housings is of critical importance because of their combined effect on the internal clearance of the bearing as well as preserving the desired degree of interference fit. The inner ring is frequently located axially by abutting against a shoulder. A fillet radius at this point is essential for the avoidance of stress concentration and the inner ring is provided with a corner radius or chamfer to allow space for this (see also Fig. 5.2a).

Where life is not the determining factor in design, it is usual to determine maximum loading by the amount to which a bearing will deflect under load. Thus the concept of "static load-carrying capacity" is understood to mean the load that can be applied to a bearing, which is either stationary or subject to slight swiveling motions, without impairing its running qualities for subsequent rotational motion. This has been determined by practical experience as the load which when applied to a bearing results in a total deformation of the rolling element and raceway at any point of contact not exceeding 0.01 per cent of the rolling-element diameter. This would correspond to a permanent deformation of 0.0025 mm for a ball 25 mm in diameter.

The successful functioning of many bearings depends upon providing them with adequate protection against their environment, and in some circumstances the environment must be protected from lubricants or products of deterioration of the bearing surfaces. Achievement of the correct functioning of seals is an essential part of bearing design. Moreover, seals which are applied to moving parts for any purpose are of interest to tribologists because they are components of bearing systems and can only be designed satisfactorily on the basis of the appropriate bearing theory.

Words and Expressions

radial bearing 向心轴承(主要用于承受径向载荷的滚动轴承)
thrust bearing 推力轴承(主要用于承受轴向载荷的滚动轴承),止推轴承

angular contact bearing 角接触轴承

tapered roller bearing 圆锥滚子轴承，也可写为 taper roller bearing

thrust ball bearing 推力球轴承(滚动体是球的推力滚动轴承)

tapered roller thrust bearing 推力圆锥滚子轴承

linear bearing 直线轴承

fivefold ['faivfəuld] *a*. *ad*. 五倍(的)，五重的

stiffness ['stifnis] *n*. 刚性，刚度，稳定性

load rating 额定载荷，额定负荷

frictional ['frikʃənl] *a*. 摩擦的，由摩擦产生的

drag [dræg] *v*. 拖，牵引，摩擦，拖着，阻碍；*n*. 阻力，摩擦力，阻尼

frictional drag 摩擦阻力

wear [wɛə] *v*. ；*n*. 磨损，磨蚀，消耗，耗损

bearing replacement 更换轴承

quantity production 大批量生产，大规模生产

contaminant [kən'tæminənt] *n*. 沾染，杂质，污染物质，污染剂

sealing [si:liŋ] *n*. 密封，封接

hostility [hɔs'tiliti] *n*. 敌视，敌意，敌对

through-harden *v*. 整体淬火，淬透

HV = Vickers hardness 维氏硬度

notwithstanding [ˌnɔtwiθ'stændiŋ] *prep*. ；*ad*. ；*conj*. 虽然，尽管……(还是)

clearance ['kliərəns] *n*. 间隙，裕度，空隙

interference [ˌintə'fiərəns] *n*. 干涉，干扰，妨碍，过盈，相互影响

interference fit 过盈配合

abut [ə'bʌt] *v*. 邻接，毗连，支撑；*n*. 端，尽头，支架

chamfer ['tʃæmfə] *n*. ；*v*. 在……开槽，倒棱，倒角，圆角

swivel ['swivl] *n*. 旋转轴承，转体；*v*. 旋转

raceway 轴承座圈，滚道

lubricant ['lju:brikənt] *n*. 润滑剂，润滑材料；*a*. 润滑的

deterioration [ditiə'reiʃən] *n*. 变质，退化，恶化，变坏

tribology [trai'bɔlədʒi] *n*. 摩擦学

12　Spindle Bearings

A machine tool spindle must provide high rotational speed, transfer torque and power to the cutting tool, and have reasonable load carrying capacity and life. The bearing system, one of the most critical components of any high-speed spindle design, must be able to meet these demands or the spindle won't perform.

High-speed spindle bearings available today include roller, tapered roller, and angular contact ball bearings (Fig. 12.1). Selection criteria depend on the spindle specifications and the speed needed for metalcutting.

(a) Roller bearing　(b) Tapered roller bearing　(c) Angular contact ball bearing

Figure 12.1　Some spindle bearings

Angular contact bearings are the type most commonly used in very high-speed spindle design. These bearings provide precision, load carrying capacity, and the speed needed for metalcutting. Angular contact ball bearings have a number of precision balls fitted into a precision steel race, and provide both axial and radial load carrying capacity when properly preloaded.

In some cases, tapered roller bearings are used because they offer higher load carrying capacity and greater stiffness than ball bearings. However, tapered roller bearings do not allow the high speeds required by many spindles.

Angular contact ball bearings are available with a choice of preloading magnitude, typically designated as light, medium, and heavy. Light preloaded bearings allow maximum speed and less stiffness. These bearings are often used for very high-speed applications, where cutting loads are light.

Heavy preloading allows less speed, but higher stiffness. To provide the required load carrying capacity for a metalcutting machine tool spindle, several angular contact ball bearings are used together. In this way, the bearings share the loads, and increase overall spindle stiffness.

Hybrid ceramic bearings are a recent development in bearing technology that uses ceramic (silicon nitride) material to make precision balls (see Figs. 12.2 and 12.3). The ceramic balls, when used in an angular contact ball bearing, offer distinct advantages over typical bearing steel balls.

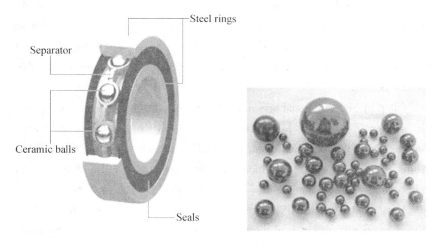

Figure 12.2　Hybrid ceramic bearing　　　Figure 12.3　Silicon nitride balls

Ceramic balls have 60% less mass than steel balls. This is significant because as a ball bearing is operating, particularly at high rotational speeds, centrifugal forces push the balls to the outer race, and even begin to deform the shape of the ball. This deformation leads to rapid wear and bearing deterioration. Ceramic balls, with less mass, will not be affected as much at the same speed. In fact, the use of ceramic balls allows up to 30% higher speed for a given ball bearing size, without sacrificing bearing life.

Due to the nearly perfect roundness of the ceramic balls, hybrid ceramic bearings operate at much lower temperatures than steel ball bearings, which results in longer life for the bearing lubricant. Tests show that spindles utilizing hybrid ceramic bearings exhibit higher rigidity and have higher natural frequencies, making them less sensitive to vibration.

Bearing lubrication is necessary for angular contact ball bearings to operate properly. The lubricant provides a microscopic film between the rolling elements to prevent abrasion. In addition, it protects surfaces from corrosion, and protects the contact area from particle contamination.

Grease is the most common and most easily applied type of lubricant. It is injected into the space between the balls and the races, and is permanent. Grease requires minimal maintenance. Generally, high-speed spindles utilizing grease lubrication do not allow replacement of the grease between bearing replacements. During a bearing replacement, clean grease is carefully injected into the bearing.

Often at high rotational speeds, lubrication with grease is not sufficient. Oil is then used as a lubricant, and delivered in a variety of ways. Maintenance of the lubrication system is vital, and must be closely monitored to ensure that proper bearing conditions are maintained. Also, use of the correct type, quantity, and cleanliness of lubricating oil is critical.

How much life can be expected? All bearings will have a useful life, defined as operation time until the bearing specifications are lost, or a complete failure of the bearing occurs. The most common cause of bearing failure is fatigue, in which the races become rough, leading to heating, and eventual mechanical failure. Bearing life, in general, is affected by axial and radial bearing loads, vibration levels, lubrication quality and quantity, maximum speed, and average bearing temperature. From these influences, bearing life can be calculated from standard equations. Today, computer models are often used to forecast bearing life.

Words and Expressions

spindle bearing 主轴轴承
rotational speed 转速(单位时间内,物体绕某轴线转动的周期数)
cutting tool 刀具
load carrying capacity 承载能力
angular contact ball bearing 角接触球轴承
roller bearing 滚子轴承
tapered roller bearing 圆锥滚子轴承
specification [ˌspesifiˈkeiʃən] *n*. 规范,说明书,[*pl*.] 技术参数,技术要求,规格
race (轴承的)滚道
preload [ˈpriːləud] *n*. 预紧,预载荷,即在施加"使用"载荷(外部载荷)前,通过相对于另一轴承的轴向调整而作用在轴承上的力,或由轴承内滚道与滚动体的尺寸形成"负游隙"(内部预载荷)而产生的力。
cutting load 切削载荷
hybrid [ˈhaibrid] *n*. 混合物; *a*. 混合的
hybrid ceramic bearing 混合陶瓷轴承(轴承钢内外圈 + 陶瓷球),复合陶瓷轴承
silicon nitride 氮化硅
roundness [ˈraundnis] *n*. 圆度
natural frequency 固有频率
rolling element 滚动体
abrasion [əˈbreiʒən] *n*. 磨损,磨耗,磨损处
bearing replacement 更换轴承
grease [griːs] *n*. 脂,润滑脂
lubrication [ˌluːbriˈkeiʃən] *n*. 润滑
cleanliness [ˈklenlinis] *n*. 清洁度
axial [ˈæksiəl] *a*. 轴的,轴向的
radial [ˈreidjəl] *a*. 径向的
standard equation 标准方程
mechanical failure 机械故障

13 Mechanisms

A mechanism has been defined as "a combination of rigid or resistant bodies so formed and connected that they move upon each other with definite relative motion."

Mechanisms form the basic geometrical elements of many mechanical devices including automatic packaging machinery, fixtures, mechanical toys, textile machinery, and others. A mechanism typically is designed to create a desired motion of a rigid body relative to a reference member. Kinematic design of mechanisms is often the first step in the design of a complete machine. When forces are considered, the additional problems of dynamics, bearing loads, stresses, lubrication, and the like are introduced, and the larger problem becomes one of machine design.

The function of a mechanism is to transmit or transform motion from one rigid body to another as part of the action of a machine. There are three types of common mechanical devices that can be used as basic elements of a mechanism.

Gear Systems Gear systems, in which toothed members in contact transmit motion between rotating shafts. Gears normally are used for the transmission of motion with a constant angular velocity ratio, although noncircular gears can be used for nonuniform transmission of motion.

Cam Systems Cam systems, where a uniform motion of an input member is converted into a nonuniform motion of the output member. The output motion may be either shaft rotation, slider translation, or other follower motions created by direct contact between the input cam shape and the follower. The kinematic design of cams involves the analytical or graphical specification of the cam surface shape required to drive the follower with a motion that is a prescribed function of the input motion.

Plane and Spatial Linkages They are also useful in creating mechanical motions for a point or rigid body. Linkages can be used for three basic tasks.

(1) Rigid body guidance. A rigid body guidance mechanism is used to guide a rigid body through a series of prescribed positions in space.

(2) Path generation. A path generation mechanism will guide a point on a rigid body through a series of points on a specified path in space.

(3) Function generation. A mechanism that creates an output motion that is a specified function of the input motion.

Mechanisms may be categorized in several different ways to emphasize their similarities and differences. One such grouping divides mechanisms into planar, spherical, and spatial categories. All three groups have many things in common; the criterion which distinguishes the groups, however, is to be found in the characteristics of the motions of the links.

A planar mechanism is one in which all particles describe plane curves in space and all these curves lie in parallel planes; i.e., the loci of all points are plane curves parallel to a single common plane. This characteristic makes it possible to represent the locus of any chosen point of a planar mechanism in its true size and shape on a single drawing or figure. The plane four-bar linkage, the plate cam and follower (see also Fig. 1.2), and the slider-crank mechanism (Fig. 13.1) are familiar examples of planar mechanisms. The vast majority of mechanisms in use today are planar.

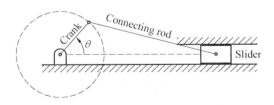

Figure 13.1 Slider-crank mechanism

A spherical mechanism is one in which each link has some point which remains stationary as the linkage moves and in which the stationary points of all links lie at a common location; i.e., the locus of each point is a curve contained in a spherical surface, and the spherical surfaces defined by several arbitrarily chosen points are all concentric. The motions of all particles can therefore be completely described by their radial projections, or "shadows," on the surface of a sphere with properly chosen center. Hooke's universal joint is perhaps the most familiar example of a spherical mechanism.

Spatial mechanisms, on the other hand, include no restrictions on the relative motions of the particles. The motion transformation is not necessarily coplanar, nor must it be concentric. A spatial mechanism may have particles with loci of double curvature. Any linkage which contains a screw pair, for example, is a spatial mechanism, since the relative motion within a screw pair is helical.

Words and Expressions

geometrical [dʒiə'metrikəl] *a.* 几何的,几何图形的
and the like = **and such like** 以及诸如此类,依此类推,等等
noncircular ['nɔn'sə:kjulə] *a.* 非圆形的
nonuniform ['nɔn'ju:nifɔ:m] *a.* 不均匀的,不一致的,非均质的
spatial ['speiʃəl] *a.* 空间的,立体的
linkage ['liŋkidʒ] *n.* 连杆机构,连接,低副运动链
prescribe [pris'kraib] *v.* 规定,命令,指示,吩咐
categorize ['kætigəraiz] *v.* 分类,把……归类,区别
criterion [krai'tiəriən] *n.* 标准,规范,准则,指标
planar ['pleinə] *a.* 平面的,在(同一)平面内的,二维的
locus ['ləukəs] (*pl. loci*) *n.* 轨迹,轨线,(空间)位置
slider-crank mechanism 曲柄滑块机构
arbitrary ['ɑ:bitrəri] *a.* 任意的,随机的,独立的,武断的
concentric [kɔn'sentrik] *a.*; *n.* 同心的,共轴的,集中的
projection [prə'dʒekʃən] *n.* 投影,射影,预测,计划
coplanar [kəu'pleinə] *a.* 共面的,同一平面的
curvature ['kə:vətʃə] *n.* 弯曲,曲率,弧度

14 Basic Concepts of Mechanisms

A combination of interrelated parts having definite motions and capable of performing useful work may be called machine. A mechanism is a component of a machine consisting of two or more bodies arranged so that the motion of one compels the motion of the others. Kinematics is the study of motion in mechanisms without reference to the forces that act on the mechanism. Dynamics is the study of motion of individual bodies and mechanisms under the influence of forces and torques. The study of forces and torques in stationary systems (and systems with negligible inertial effects) is called statics.

The definitions of some terms and concepts fundamental to the study of kinematics and dynamics of mechanisms are defined below.

Degrees of Freedom Any mechanical system can be classified according to the number of degrees of freedom (DOF) which it possesses. The system's DOF is equal to the number independent parameters which are needed to uniquely define its position in space at any instant time.

An unconstrained rigid body has six degrees of freedom: translation in three coordinate directions and rotation about three coordinate axes. If the body is restricted to motion in a plane, there are three degrees of freedom: translation in two coordinate directions and rotation within the plane.

Kinematic Pair The connections between two or more links that permit constrained relative motion are called kinematic pairs.

Lower and Higher Pairs Connections between rigid bodies consist of lower and higher pairs of elements. The two elements of a lower pair have theoretical surface contact with one another, while the two elements of a higher pair have theoretical point or line contact (if we disregard deflections).

Link A link is one of the rigid bodies or members joined together to form a kinematic chain. The term rigid link, or sometimes simply link, is an idealization used in the study of mechanisms that does not consider small deflections due to strains in machine members. A perfectly rigid or inextensible link can exist only as a textbook type of model of a real machine member. For typical machine parts, the maximum changes in dimension are on the order of only a one-thousandth of the part length. We are justified in neglecting this small motion when considering the much greater motion characteristic of most mechanisms. The word link is used in a general sense to include cams, gears, and other machine members in addition to cranks, connecting rods and other pin-connected components.

Frame The fixed or stationary link in a mechanism is called the frame. When there is no link that is actually fixed, we may consider one as being fixed and determine the motion of the other links relative to it. In an automotive engine, for example, the engine block (see Fig. 14.1) is considered the frame, even though the automobile may be moving.

Figure 14.1   Engine block

Linkage   Since we may wish to examine kinematic chains without regard to their ultimate use, it is convenient to identify any assemblage of rigid bodies connected by kinematic pairs as a linkage. Thus, both mechanisms and machines may be considered linkages.

The number of degrees of freedom of a linkage is the number of independent parameters required to specify the position of every link relative to the frame or fixed link. There is no requirement that a mechanism have only one DOF, although this is often desirable for simplicity. Some machines have many DOF.

Kinematic Chain A kinematic chain is an assembly of links and kinematic pairs. kinematic chains may be either open or closed (Fig. 14.2). Each link in a closed kinematic chain is connected to two or more other links. A linkage failing to meet this criterion is an open kinematic chain. In this case, one (or more) of the links is connected to one other link. The industrial robot shown in Fig.53.1 is an open kinematic chain.

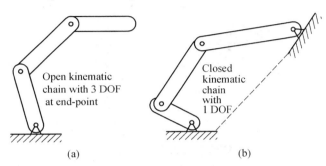

(a) (b)

Figure 14.2 Examples of open (a) and closed (b) kinematic chains

Planar Motion and Planar Linkages If all points in a linkage move in parallel planes, the system undergoes planar motion and the linkage may be described as a planar linkage. If we examine the linkage made up of the crank, connecting rod, and piston in a piston engine, we see that it is a planar linkage. Most of the mechanisms in common use may be treated as planar linkages.

Spatial Motion and Spatial Linkages The more general case in which motion cannot be described as taking place in parallel planes is called spatial motion, and the linkage may be described as a spatial or three-dimensional (3D) linkage.

Kinematic Inversion The absolute motion of a linkage depends on which of the links is fixed, that is, which link is selected as the frame. If two otherwise equivalent linkages have different fixed link, then each is an inversion of the other. Figure 14.3 shows the four possible inversions of a four-bar linkage: two crank-rocker linkages (Figs. 14.3a and 14.3b), a double-rocker linkage (Fig. 14.3c), and a double-crank linkage (Fig. 14.3d).

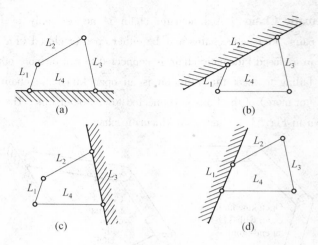

Figure 14.3 Kinematic inversions of a four-bar linkage

Words and Expressions

compel [kəm'pel] *v*. 强迫,迫使
kinematics [ˌkaini'mætiks] *n*. 运动学
degree of freedom 自由度
unconstrained ['ʌnkən'streind] *a*. 不受拘束的,自由的,无约束的
translation [træns'leiʃən] *n*. 平移,移动
coordinate direction 坐标方向
kinematic pair 运动副
link [liŋk] *n*. 构件
lower and higher pairs 低副和高副
kinematic chain 运动链(用运动副连接而成的相对可动的构件系统)
inextensible [ˌiniks'tensəbl] *a*. 不能拉长的,不能伸展的
pin-connected 用销连接的,铰接的
frame [freim] *n*. 机架,固定构件(fixed link)
engine block 发动机缸体
closed kinematic chain 闭式运动链(每个构件至少与两个其他构件以运动副相连接的运动链)
open kinematic chain 开式运动链(运动链中至少有一处未形成闭环的运动链)

planar motion 平面运动
kinematic inversion 机架变换
inversion [inˈvəːʃən] ***n***. 倒置,机架变换,变换
crank-rocker linkage 曲柄摇杆机构
double-rocker linkage 双摇杆机构
double-crank linkage 双曲柄机构

15 Introduction to Tribology

Tribology is defined as the science and technology of interacting surfaces in relative motion, having its origin in the Greek word tribos meaning rubbing. It is a study of the friction, lubrication, and wear of engineering surfaces with a view to understanding surface interactions in detail and then prescribing improvements in given applications. The work of the tribologist is truly interdisciplinary, embodying physics, chemistry, mechanics, thermodynamics, and materials science, and encompassing a large, complex, and interwinded area of machine design, reliability, and performance where relative motion between surfaces is involved.

It is estimated that approximately one-third of the world's energy resources in present use appear as friction in one form or another. This represents a staggering loss of potential power for today's mechanized society. The purpose of research in tribology is understandably the minimization and elimination of unnecessary waste at all levels of technology where the rubbing of surfaces is involved.

One of the important objectives in tribology is the regulation of the magnitude of frictional forces according to whether we require a minimum (as in machinery) or a maximum (as in the case of anti-skid surfaces). It must be emphasized, however, that this objective can be realized only after a fundamental understanding of the frictional process is obtained for all conditions of temperature, sliding velocity, lubrication, surface roughness, and material properties.

The most important criterion from a design viewpoint in a given application is whether dry or lubricated conditions are to prevail at the sliding interface. In many applications such as machinery, it is known that only one condition shall prevail (usually lubrication), although several regimes of lubrication may exist. There are a few cases, however, where it cannot be known in advance whether the interface is dry or wet, and it is obviously more difficult to proceed with any design. The commonest example of this phenomenon is the pneumatic tyre. Under dry conditions it is desirable to maximize the adhesion component of friction by ensuring a maximum contact area between tyre and road—and this is achieved by having a smooth tread and a smooth road surface. Such a combination, however, would produce a disastrously low coefficient of friction under wet conditions. In the latter case, an adequate tread pattern and a suitably textured road surface offer the best conditions, although this combination gives a lower coefficient of friction in dry weather.

The several lubrication regimes which exist may be classified as hydrodynamic, boundary, and elastohydrodynamic. The different types of bearing used today are the best examples of fully hydrodynamic behaviour, where the sliding surfaces are completely separated by an interfacial lubricant film. Boundary or mixed lubrication is a combination of hydrodynamic and solid contact between moving surfaces, and this regime is normally assumed to prevail when hydrodynamic lubrication fails in a given product design. For example, a journal bearing is designed to operate at a given load and speed in the fully hydrodynamic region, but a fall in speed or an increase in load may cause part solid and part hydrodynamic lubrication conditions to occur between the journal and bearing surfaces. This boundary lubrication condition is unstable, and normally recovers to the fully hydrodynamic behaviour or degenerates into complete seizure of the surfaces. The pressures developed in thin lubricant films may reach proportions capable of elastically deforming the boundary surfaces of the lubricant, and conditions at the sliding interface are then classified as elastohydrodynamic.

It is now generally accepted that elastohydrodynamic contact conditions exist in a variety of applications hitherto considered loosely as belonging to the hydrodynamic or boundary lubrication regimes; for example, the contact of mating gear teeth. Solid lubricants exhibit a compromise between dry and lubricated conditions in the sense that although the contact interface is normally dry, the solid lubricant material behaves as though initially wetted. This is a consequence of a physico-chemical interaction occurring at the surface of a solid lubricant lining under particular loading and sliding conditions, and these produce the equivalent of a lubricating effect.

Words and Expressions

interdisciplinary 各学科之间的,边缘学科的,多种学科的
thermodynamics [ˈθəːməudaiˈnæmiks] *n*. 热力学
encompass [inˈkʌmpəs] *v*. 围绕,包含,包括,拥有,完成
interwind [ˌintəˈwaind] *v*. 互相盘绕,互卷
staggering [ˈstæɡəriŋ] *n*. 交错; *a*. 交错的,惊人的,压倒的
antiskid [ˈæntiˈskid] *n*.; *a*. 防滑的,防滑轮胎纹
prevail [priˈveil] *v*. 流行,普及,占优势,经常发生
interface [ˌintəˈfeis] *n*. 交界面,分界面,界面,相互作用面
interfacial [ˌintəˈfeiʃəl] *a*. 分界面的,两表面间的,界面的
proceed [prəˈsiːd] *v*. 进行,继续做下去,开始,着手
pneumatic [njuːˈmætik] *a*. 气动的,气压的,充气的
regime [reiˈʒiːm] *n*. 状况,状态,方式,方法
tread [tred] *n*. 踩,踏,滑动面,轮胎花纹
disastrously [diˈzɑːstrəsli] *ad*. 灾害性地,造成巨大损失地
boundary lubrication 边界润滑
hydrodynamic [ˈhaidrəudaiˈnæmik] *a*. 流体的,流体动力(学)的
elastohydrodynamics 弹性流体动力学
degenerate [diˈdʒenəreit] *v*. 退化,变质,简并
seizure [ˈsiːʒə] *n*. 轧住,咬住,卡住,塞住,咬缸
hitherto [ˈhiðətuː] *ad*. 迄今,至今,向来,从来
loosely [ˈluːsli] *ad*. 松散地,不精确地,不严格地,大概
compromise [ˈkɔmprəmaiz] *n*.; *v*. 妥协,折衷,兼顾,综合考虑
behave as 性能[作用,表现]像……一样
lining [ˈlainiŋ] *n*. 衬层,涂层,覆盖

16 Wear and Lubrication

Wear can be defined as the progressive loss of material from the operating surface of a body occurring as a result of relative motion at the surface. The problem of wear arises wherever there are load and motion between surfaces, and is therefore important in engineering practice, often being the major factor governing the life and performance of machine components. The major types of wear are described next:

Adhesive Wear. When two surfaces are loaded against each other, the whole of the contact load is carried on very small area of the asperity contacts. The real contact pressure at these asperities is very high, and adhesion takes place between them. If a tangential force is applied to the model shown in Fig. 16.1, shearing can take place either (a) at the interface or (b) below or above the interface, causing adhesive wear. Because of factors such as strain hardening at the asperity contact, the adhesive bonds often are stronger than the base metals. Thus, during sliding, fracture usually occurs in the weaker or softer component, and a wear fragment is generated. Although this fragment is attached to the harder component(upper surface in Fig. 16.1c), it eventually becomes detached during further rubbing at the interface and develops into a loose wear particle.

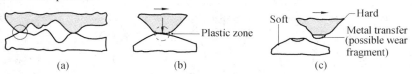

Figure 16.1　Schematic illustration of (a) two contacting asperities, (b) adhesive between two asperities, and (c) the formation of a wear particle

Abrasive Wear. This type of wear is caused by a hard, rough surface (or a surface containing hard, protruding particles) sliding across another surface. As a result, microchips are produced, thereby leaving grooves or scratches on the softer surface (Fig. 16.2). In fact, processes such as filing and grinding act in this manner. The difference is that, in these operations, the process parameters are controlled to produce the desired shapes and surfaces through wear;whereas, abrasive wear generally is unintended and unwanted.

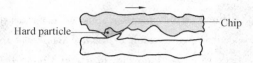

Figure 16.2 Schematic illustration of abrasive wear in sliding

Corrosive Wear. Also known as oxidation or chemical wear, this type of wear is caused by chemical or electrochemical reactions between the surfaces and the environment. The fine corrosive products on the surface constitute the wear particles in this type of wear. When the corrosive layer is destroyed or removed through sliding, another layer begins to form, and the process of removal and corrosive layer formation is repeated.

Fatigue Wear. Fatigue wear is caused when the surface of a material is subjected to cyclic loading (see Fig. 16.3); one example of this is the rolling contact in bearings. The wear particles usually are formed through spalling or pitting. Another type of fatigue wear is by thermal fatigue. Cracks on the surface are generated by thermal stresses from thermal cycling, such as when a cool forging die repeatedly contacts hot workpieces. These cracks then join, and the surface begins to spall, producing fatigue wear.

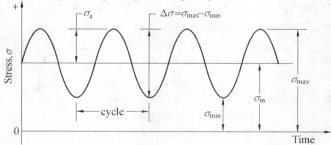

Figure 16.3 Cyclic loading parameters

Although wear generally alters a part's surface topography and may result in severe surface damage, it also can have a beneficial effect. The running-in period for various machines and engines produces this type of wear by removing the peaks from asperities (Fig. 16.4). Thus, under controlled conditions, wear may be regarded as a type of smoothing or polishing process.

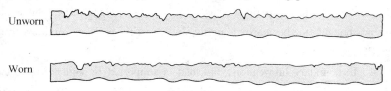

Figure 16.4　Changes in original ground-surface profiles after wear

Lubrication is a process by which the friction and wear between two solid surfaces in relative motion is reduced significantly by the interposing a lubricant between them. There are four types of fluid lubrication that are generally of interest in engineering practice (Fig. 16.5):

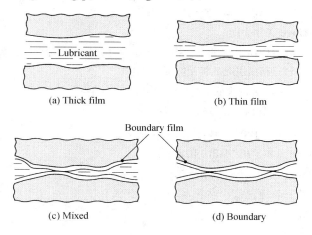

Figure 16.5　Types of lubrication

1. In thick-film lubrication, the two surfaces are completely separated from each other by a continuous fluid film. The fluid film can be developed either hydrostatically or, more often by the wedge effect of the sliding surfaces

· 63 ·

in the presence of a viscous fluid at the interface.

2. As the load between the two surfaces increases or as the sliding speed and viscosity of the fluid decrease, the lubricant film becomes thinner (thin-film lubrication).

3. In mixed lubrication, although the contacting surfaces are separated by a thin lubricant film, asperity contact also may take place. The total applied load is thought to be carried partly by asperity contacts and partly by hydrodynamic action.

4. In boundary lubrication, the load is supported by contacting surfaces covered with a boundary film of lubricant (Fig. 16.5d). This is a thin (molecular) lubricant layer that is attracted physically to the metal surfaces, thus preventing direct metal-to-metal contact of the two bodies and reducing wear.

Words and Expressions

progressive [prə'gresiv] *a*. 逐渐的,逐步的
operating surface 工作表面
adhesive wear 粘着磨损,黏着磨损,黏附磨损(由于黏附作用使两摩擦表面的材料迁移而引起的机械磨损)
asperity [æs'periti] *n*. (表面上的)微小凸出物,凸峰
interface ['intə(:)ˌfeis] *n*. 结合面,接触面
tangential [tæn'dʒenʃ(ə)l] *a*. 切线的,切向的
adhesion [əd'hi:ʒən] *n*. 粘着,附着,黏着
strain hardening 应变硬化,加工硬化
adhesive bond 粘着键,粘结,粘着力
base metal 基底金属,基体金属
metal transfer 金属材料转移
wear fragment 由磨损产生的碎片
abrasive wear 磨料磨损,磨粒磨损
microchip ['maikrəʊtʃip] *n*. 细小的切屑
filing ['failiŋ] *n*. 锉削,用锉刀锉

grinding ['graindiŋ] *n*. 磨削
unintended ['ʌnin'tendid] *a*. 非计划中的,非故意的, 无意识的
corrosive wear 腐蚀磨损
electrochemical reaction 电化学反应
fine corrosive product 细小的腐蚀生成物
fatigue wear 疲劳磨损
cyclic loading 交变载荷,循环载荷(周期性或非周期性经一定时间后重复出现的动载荷)
spalling ['spɔ:liŋ] *n*. 剥落(疲劳磨损时从摩擦表面以鳞片形式分离出磨屑的现象)
pitting ['pitŋ] *n*. 点蚀,剥蚀(疲劳磨损时从摩擦表面以颗粒形式分离出磨屑,并在摩擦表面留下"痘斑"的磨损)
thermal stress 热应力
forging die 锻模
topography [tə'pɔgrəfi] *n*. 形貌
surface topography 表面形貌
running-in period 走合期,磨合期
controlled condition 受控条件,一定条件
polishing ['pɔliʃiŋ] *n*. 抛光(使工件获得光亮、平整表面的加工方法)
worn [wɔ:n] *a*. 用旧的,磨损的, wear 的过去分词
unworn ['ʌn'wɔ:n] *a*. 没有磨损的
ground surface 磨削表面,经过磨削加工后的表面
boundary film 边界膜
thick-film lubrication 厚膜润滑(在工作载荷下润滑膜厚度远大于表面微凸体高度,无表面粗糙度效应的润滑)
wedge effect 楔效应(黏性流体按收敛方向流入楔形间隙而产生压力的效应)
thin-film lubrication 薄膜润滑
mixed lubrication 混合润滑
boundary lubrication 边界润滑
physically ['fizik(ə)li] *ad*. 物理地

17 Fundamentals of Mechanical Design

Mechanical design means the design of things and systems of a mechanical nature—machines, products, structures, devices, and instruments. For the most part mechanical design utilizes mathematics, the materials sciences, and the engineering mechanics.

The total design process is of interest to us. How does it begin? Does the engineer simply sit down at his desk with a blank sheet of paper? And, as he jots down some ideas, what happens next? What factors influence or control the decisions which have to be made? Finally, how does this design process end?

Sometimes, but not always, design begins when an engineer recognizes a need and decides to do something about it. Recognition of the need and phrasing it in so many words often constitute a highly creative act because the need may be only a vague discontent, a feeling of uneasiness, or a sensing that something is not right.

The need is usually not evident at all. For example, the need to do something about a food-packaging machine may be indicated by the noise level, by the variation in package weight, and by slight but perceptible variations in the quality of the packaging.

There is a distinct difference between the statement of the need and the identification of the problem which follows this statement. The problem is more specific. If the need is for cleaner air, the problem might be that of reducing the dust discharge from power plant stacks, or reducing the quantity of irritants from automotive exhausts (see Figs. 17.1 and 17.2).

Definition of the problem must include all the specifications for the thing that is to be designed. The specifications are the input and output quantities, the characteristics and dimensions of the space the thing must occupy and all the limitations on these quantities. The specifications define the cost, the number to be manufactured, the expected life, the operating temperature, and the reliability.

Figure 17.1   Power plant stacks

Figure 17.2   Irritants from automobile exhausts

There are many implied specifications which result either from the designer's particular environment or from the nature of the problem itself. The manufacturing processes which are available, together with the facilities of a certain plant, constitute restrictions on a designer's freedom, and hence are a part of the implied specifications. A small plant, for instance, may not own cold-working machinery. Knowing this, the designer selects other metal-processing methods which can be performed in the plant. The labor skills available and the competitive situation also constitute implied specifications.

After the problem has been defined and a set of written and implied specifications has been obtained, the next step in design is the synthesis of an optimum solution. Now synthesis cannot take place without both analysis and optimization because the system under design must be analyzed to determine whether the performance complies with the specifications.

The design is an iterative process in which we proceed through several steps, evaluate the results, and then return to an earlier phase of the procedure. Thus we may synthesize several components of a system, analyze and optimize them, and return to synthesis to see what effect this has on the remaining parts of the system. Both analysis and optimization require that we construct or devise abstract models of the system which will admit some form of mathematical analysis. We call these models mathematical models. In creating them it is our hope that we can find one which will simulate the real physical system very well.

Evaluation is a significant phase of the total design process. Evaluation is the final proof of a successful design, which usually involves the testing of a prototype in the laboratory. Here we wish to discover if the design really satisfies the need or needs. Is it reliable? Will it compete successfully with similar products? Is it economical to manufacture and to use? Is it easily maintained and adjusted? Can a profit be made from its sale or use?

Communicating the design to others is the final, vital step in the design process. Undoubtedly many great designs, inventions, and creative works have been lost to mankind simply because the originators were unable or unwilling to explain their accomplishments to others. Presentation is a selling job. The engineer, when presenting a new solution to administrative, management, or supervisory persons, is attempting to sell or to prove to them that this solution is a better one. Unless this can be done successfully, the time and effort spent on obtaining the solution have been largely wasted.

Basically, there are only three means of communication available to us. These are the written, the oral, and the graphical forms. Therefore the successful engineer will be technically competent and versatile in all three forms of communication. A technically competent person who lacks ability in any one of these forms is severely handicapped. If ability in all three forms is lacking, no one will ever know how competent that person is!

The competent engineer should not be afraid of the possibility of not succeeding in a presentation. In fact, occasional failure should be expected because failure or criticism seems to accompany every really creative idea. There is a great deal to be learned from a failure, and the greatest gains are obtained by those willing to risk defeat. In the final analysis, the real failure would lie in deciding not to make the presentation at all.

Words and Expressions

for the most part 大部分,在大多数情况下
blank [blæŋk] *a.*; *n.* 空白,空页,坯料
jot [dʒɔt] *v.* 把……摘记下来,匆匆地记下; *n.* 一点,少许,小额
recognition [ˌrekəg'niʃən] *n.* 认识,识别,辨别,承认,重视,认可
vague [veig] *a.* 不明确的,含糊的,未定的,不明的
discontent ['diskən'tent] *n.* 不满意; *a.* 不满的,不安的; *v.* (令人)不满
perceptible [pə'septibl] *a.* 可感觉到的,能觉察得出的,明显的
irritant ['iritənt] *a.* 刺激的,有刺激性的; *n.* 刺激物
exhaust [ig'zɔ:st] *v.* 用尽,排出,排气; *n.* 排气,排气装置
implied [im'plaid] *a.* 暗指的,含蓄的,不言而喻的
synthesis ['sinθisiis] *n.* 合成,综合,结构综合
optimum ['ɔptiməm] *a.*; *n.* 最佳(的,条件,方式),最优的,最有利的
optimize ['ɔptimaiz] *v.* 优选,选择最佳条件,发挥最大作用
comply [kəm'plai] *v.* 答应,同意,遵守,履行,根据
synthesize ['sinθisaiz] *v.* 合成,综合,接合
devise [di'vaiz] *v.* 设计,计划,发明,创造,产生
prototype ['prəutətaip] *n.* 原型,样机,模型机
originator [ə'ridʒineitə] *n.* 创作者,发明者,创办人,发起人
accomplishment [ə'kɔmpliʃmənt] *n.* 完成,实施,成就,成绩,本领,技能
presentation [ˌprezen'teiʃən] *n.* 提出,展示,表示,表现
versatile ['vəsətail] *a.* 通用的,多用途的,多方面的
handicap ['hændikæp] *n.* 障碍,不利条件,缺陷; *v.* 妨碍,为……的障碍
in the final analysis 总之,归根到底

18 Machine Designer's Responsibility

A new machine is born because there is a real or imagined need for it. It evolves from someone's conception of a device with which to accomplish a particular purpose. From the conception follows a study of the arrangement of the parts, the location and length of links (which may include a kinematic study of the linkage), the places for gears, bolts, springs, cams, and other elements of machines. Detailed analyses of displacements, velocities, and accelerations are usually required. This part of the design process is followed by an analysis of forces and torques. With all ideas subject to change and improvement, several solutions may be and usually are found, the seemingly best one being chosen.

The actual practice of designing is applying a combination of scientific principles and a knowing judgment based on experience. It is seldom that a design problem has only one right answer, a situation that is often annoying to the beginner in machine design.

Design problems usually have more than one answer. Given a general statement of a design problem, such as design a machine to wash clothes in the home automatically, and there will be as many different answers as there are design teams—as attested by the number of washing machines (see Fig. 18.1) on the market.

In order to successfully compete from year to year, most manufacturers must continuously modify their product and their methods of production. Increases in the production rate, upgrading of product performance, redesign for cost and weight reduction are frequently required.

Engineering practice usually requires compromises. Competition may require a reluctant decision contrary to one's best engineering judgment; production difficulties may force a change of design; etc.

Figure 18.1　Automatic washing machines

A good designer needs many attributes, for example:

(1) A good background in strength of materials, so that the stress analyses are sound. The parts of the machine should have adequate strength and rigidity, or other characteristics as needed.

(2) A good acquaintance with the properties of materials used in machines.

(3) A familiarity with the major characteristics and economics of various manufacturing processes, because the parts that make up the machine must be manufactured at a competitive cost. It happens that a design that is economic for one manufacturing plant may not be so for another. For example, a plant with a well-developed welding department but no foundry might find that welding is the most economic fabricating method in a particular situation; whereas another plant faced with the same problem might decide upon casting because they have a foundry (and may or may not have a welding department).

(4) A specialized knowledge in various circumstances, such as the properties of materials in corrosive atmospheres, at very low (cryogenic) temperatures, or at relatively high temperatures.

(5) A preparation for deciding wisely: (a) when to use manufacturers' catalogs, buying stock or relatively available items, and when custom design is necessary; (b) when empirical design is justified; (c) when the design should be tested in service tests before manufacture starts; (d) when special measures

should be taken to control vibration and sound (and others).

(6) Some aesthetic sense, because the product must have "customer appeal" if it is to sell.

(7) A knowledge of economics and comparative costs, because the best reason for the existence of engineers is that they save money for those who employ them. Anything that increases the cost should be justified by, for instance, an improvement in performance, the addition of an attractive feature, or greater durability.

(8) Inventiveness and the creative instinct, most important of all for maximum effectiveness. Creativeness may arise because an energetic mind is dissatisfied with something as it is and this mind is willing to act.

Naturally, there are many other important considerations and a host of details. Will the machine be safe to operate? Is the operator protected from his own mistakes and carelessness? Is vibration likely to cause trouble? Will the machine be too noisy? Is the assembly of the parts relatively simple? Will the machine be easy to service and repair?

Of course, no one engineer is likely to have enough expert knowledge concerning the above attributes to make optimum decisions on every question. The larger organizations will have specialists to perform certain functions, and smaller ones can employ consultants. Nevertheless, the more any one engineer knows about all phases of design, the better. Design is an exacting profession, but highly fascinating when practiced against a broad background of knowledge.

Words and Expressions

conception [kən'sepʃən] *n*. 构思,构想,概念,观念
arrangement [ə'reindʒmənt] *n*. 配置,布局,构造,方案,装置
seemingly ['si:miŋli] *ad*. 表面上,外观上,看上去
knowing ['nəuiŋ] *a*. 会意的,显示出聪明的洞察力和智慧的,精明的
attest [ə'test] *v*. 证明,证实,表明,郑重宣布,为……作证
reluctant [ri'lʌktənt] *a*. 不愿的,勉强的,难得到的
attribute ['ætribju:t] *n*. 属性,特性,特征,标志,象征

sound [saund] *n.* 声音, *a.* 完整的, 正确的, 合理的, 有根据的
acquaintance [əˈkweintəns] *n.* 熟悉, 了解, 相识, 感性认识, 心得
well-developed 发达的, 良好的
foundry [ˈfaundri] *n.* 铸造厂, 铸造车间
cryogenic [ˌkraiəˈdʒenik] *a.* 冷冻的, 低温的, 制冷的, 深冷的
catalog [ˈkætələg] *n.* 目录, 种类, (产品)样本
stock [stɔk] *n.* 原料, 材料, 坯料, 毛坯, 存货
custom design 定制设计, 用户设计, 按用户需求设计
empirical [emˈpirikəl] *a.* 经验的, 实验的, 以实验为根据的
aesthetic [iːsˈθetik] *a.* 审美的, 美学的, 美术的
comparative cost 比较成本(指对企业的决策有重要意义的各种成本的比较。诸如, 直接成本与间接成本, 制造成本与非制造成本等的比较)
inventiveness [inˈventivnis] *n.* 发明创造能力, 创造性
instinct [ˈinstiŋkt] *n.* 本性, 本能, 直觉
effectiveness *n.* 效率, 效果, 效能, 有效性, 功效
energetic [ˌenəˈdʒetik] *a.* 能的, 有力的, 精力旺盛的
as it is 按照原状, 实际上
host [həust] *n.* 许多, 多数
a host of 许多, 一大群, 一大批
service [ˈsəːvis] *n.*; *v.* 服务, 运转, 使用, 操作, 维修
consultant [kənˈsʌltənt] *n.* 顾问, 咨询, 商议者
fascinating [ˈfæsineitiŋ] *a.* 引人入胜的, 极有趣的
exacting [igˈzæktiŋ] *a.* 严格的, 苛求的, 艰难的, 需付出极大努力的

19 Introduction to Machine Design

Machine design is the application of science and technology to devise new or improved products for the purpose of satisfying human needs. It is a vast field of engineering technology which not only concerns itself with the original conception of the product in terms of its size, shape and construction details, but also considers the various factors involved in the manufacture, marketing and use of the product.

People who perform the various functions of machine design are typically called designers, or design engineers. Machine design is basically a creative activity. However, in addition to being innovative, a design engineer must also have a solid background in the areas of mechanical drawing, kinematics, dynamics, materials engineering, strength of materials and manufacturing processes.

As stated previously, the purpose of machine design is to produce a product which will serve a need for man. Inventions, discoveries and scientific knowledge by themselves do not necessarily benefit people; only if they are incorporated into a designed product will a benefit be derived. It should be recognized, therefore, that a human need must be identified before a particular product is designed.

Machine design should be considered to be an opportunity to use innovative talents to envision a design of a product, to analyze the system and then make sound judgments on how the product is to be manufactured. It is important to understand the fundamentals of engineering rather than memorize mere facts and equations. There are no facts or equations which alone can be used to provide all the correct decisions required to produce a good design. On the other hand, any calculations made must be done with the utmost care and precision. For example, if a decimal point is misplaced, an otherwise acceptable design may not function.

Good designs require trying new ideas and being willing to take a certain amount of risk, knowing that if the new idea does not work the existing method can be reinstated. Thus a designer must have patience, since there is no assurance of success for the time and effort expended. Creating a completely new design generally requires that many old and well-established methods be thrust aside. This is not easy since many people cling to familiar ideas, techniques and attitudes. A design engineer should constantly search for ways to improve an existing product and must decide what old, proven concepts should be used and what new, untried ideas should be incorporated.

New designs generally have "bugs" or unforeseen problems which must be worked out before the superior characteristics of the new designs can be enjoyed. Thus there is a chance for a superior product, but only at higher risk. It should be emphasized that, if a design does not warrant radical new methods, such methods should not be applied merely for the sake of change.

During the beginning stages of design, creativity should be allowed to flourish without a great number of constraints. Even though many impractical ideas may arise, it is usually easy to eliminate them in the early stages of design before firm details are required by manufacturing. In this way, innovative ideas are not inhibited. Quite often, more than one design is developed, up to the point where they can be compared against each other. It is entirely possible that the design which is ultimately accepted will use ideas existing in one of the rejected designs that did not show as much overall promise.

Psychologists frequently talk about trying to fit people to the machines they operate. It is essentially the responsibility of the design engineer to strive to fit machines to people. This is not an easy task, since there is really no average person for which certain operating dimensions and procedures are optimum.

Another important point which should be recognized is that a design engineer must be able to communicate ideas to other people if they are to be incorporated. Initially, the designer must communicate a preliminary design to get management approval. This is usually done by verbal discussions in conjunction with drawing layouts and written material. To communicate effectively, the following questions must be answered:

(1) Does the design really serve a human need?
(2) Will it be competitive with existing products of rival companies?
(3) Is it economical to produce?
(4) Can it be readily maintained?
(5) Will it sell and make a profit?

Only time will provide the true answers to the preceding questions, but the product should be designed, manufactured and marketed only with initial affirmative answers. The design engineer also must communicate the finalized design to manufacturing through the use of detail and assembly drawings.

Quite often, a problem will occur during the manufacturing cycle. It may be that a change is required in the dimensioning or tolerancing of a part so that it can be more readily produced. This falls in the category of engineering changes which must be approved by the design engineer so that the product function will not be adversely affected. In other cases, a deficiency in the design may appear during assembly or testing just prior to shipping. These realities simply bear out the fact that design is a living process. There is always a better way to do it and the designer should constantly strive towards finding that better way.

Words and Expressions

envision [en′viʒən] *v.* 想象,预见,展望
innovative [′inəuveitiv] *a.* 革新的,创新的
mere [miə] *a.* 仅仅的,只不过; *n.* 边界
utmost [′ʌtməust] *a.*; *n.* 极度(的),极端(的),最大限度(的),最大可能
decimal [′desiməl] *a.* 十进制的,小数的; *n.* (十进制)小数
decimal point 小数点
misplace [mis′pleis] *v.* 错放,误置,放错位置

reinstate [´ri:in´steit] *v*. 复原,修复,恢复,使正常,使恢复原状
well-established 固定下来的,得到确认的,良好的
cling [kliŋ] *v*. 粘着,坚持,依附于,依靠,抱着……不放
bug [bʌg] *n*. 缺陷,故障,障碍,困难
unforeseen [´ʌnfɔ:´si:n] *a*. 想不到的,未预见到的,偶然的
superior [sju:´piəriə] *a*. 高级的,优良的,上级的,占优势的;*n*. 优胜者
flourish [´flʌriʃ] *v*.;*n*. 繁荣,兴旺,蓬勃发展
inhibit [in´hibit] *v*. 防止,阻止,抑制,防腐蚀
detail drawing 零件图
assembly drawing 装配图
psychologist [sai´kɔlədʒist] *n*. 心理学家
fall in 落入,符合,归于,属于
incorporate [in´kɔ:pəreit] *v*. 结合,合并,组成公司
affirmative [ə´fə:mətiv] *a*. 肯定的,正面的,赞成的;*n*. 肯定,确认
deficiency [di´fiʃənsi] *n*. 缺乏,不足之处,缺陷,不足额,亏空
bear out 证明,证实

20   Some Rules for Mechanical Design

The old saying that "necessity is the mother of invention" is still true, and a new machine, structure, or device is the result of that need. If the new machine is really needed or desired, people will buy it or use it as long as they can afford it.

Before a new machine of any kind goes into production, certain questions must be answered: What is the potential market for this machine? Can the machine be sold at a price that people are willing to pay? If the potential market is large enough and the estimated selling price seems reasonable, then the inventor, designer, or company officials may choose to proceed with the development, production, and marketing plans for the new project.

A new machine may be needed to perform a function previously done by men, such as automated assembly. An existing machine may need improvements in durability, efficiency, weight, speed, or cost. With the objective wholly or partly defined, the next step in design is the conception of mechanisms and their arrangements that will perform the needed functions. For this, freehand sketching is of great value, not only as a record of one's thoughts and as an aid in discussion with others, but particularly for communication with one's own mind, as a stimulant for creative ideas.

When the general shape and a few dimensions of the several components become apparent, analysis can begin in earnest. The analysis will have as its objective satisfactory or superior performance, plus safety and durability with minimum weight, and a competitive cost. Optimum proportions and dimensions will be sought for each critically loaded section, together with a balance between the strength of the several components. Materials and their treatment will be chosen. These important objectives can be attained only by analysis based upon the principles of mechanics, such as those of statics for reaction forces and for the optimum utilization of friction; of dynamics for inertia, acceleration, and energy; of elasticity and strength of materials for stress and deflection; and of physical behavior of materials.

Finally, a design based upon function and reliability will be completed, and a prototype may be built. If its tests are satisfactory, and if the device is to be produced in quantity, the initial design will undergo certain modifications that enable it to be manufactured in quantity at a lower cost. During subsequent years of manufacture and service, the design is likely to undergo changes as new ideas are conceived or as further analysis based upon tests and experience indicate alterations. Market demand, customer satisfaction, and manufacture cost are all related to design, and ability in design is intimately involved in the success of technology innovation.

To stimulate creative thought, the following rules are suggested for the designer.

1. Utilize desired physical properties and to control undesired ones

The performance requirements of a machine are met by utilizing laws of nature or properties of matter (e. g., flexibility, strength, gravity, inertia, buoyancy, centrifugal force, principles of the lever and inclined plane, friction, viscosity, fluid pressure, and thermal expansion), also the many electrical, optical, thermal, and chemical phenomena. However, what may be useful in one application may be detrimental in the next. Friction is desired at the clutch facing (see Figs. 20.1 and 20.2) but not in the clutch bearing. Ingenuity in design should be applied to utilize and control the physical properties that are desired and to minimize those that are not desired.

2. Provide for favorable stress distribute and stiffness

On components subjected to fluctuating stress, particular attention is given to a reduction in stress concentration, and to an increase of strength at fillets (see Figs. 5.2 and 39.1a), threads, and holes. Stress reductions are made by modification in shape. Hollow shafts (see also Fig. 6.3) and box sections give a favorable stress distribution, together with stiffness and minimum weight. The stiffness of shafts and other components must be suitable to avoid resonant vibrations.

Figure 20.1　A clutch　　　　Figure 20.2　Clutch facings

3. Use basic equations to calculate and optimize dimensions

The fundamental equations of mechanics and the other sciences are the accepted bases for calculations. They are sometimes rearranged in special forms to facilitate the determination or optimization of dimensions, such as the beam and surface stress equations for determining gear-tooth size. Factors may be added to a fundamental equation for conditions not analytically determinable, e.g., on thin steel tubes, an allowance for corrosion added to the thickness based on pressure.

4. Choose materials for a combination of properties

Materials should be chosen for a combination of pertinent properties, not only for strengths, hardness, and weight, but sometimes for resistance to impact, corrosion, and low or high temperatures. Cost and fabrication properties are factors, such as weldability, machinability, sensitivity to variation in heat-treating temperatures, and required coating.

5. Provide for accurate location and noninterference of parts in assembly

A good design provides for the correct locating of parts and for easy assembly and repair. Shoulders give accurate location without measurement during assembly. Shapes can be designed so that parts cannot be assembled backwards or in the wrong place. Interferences, as between screws in tapped holes, and between linkages must be foreseen and prevented.

Words and Expressions

selling price 销售价格
marketing plan 市场销售计划
automated assembly 自动装配
conception [kən'sepʃən] *n.* 构思，构想，设想，见解，想法
freehand sketching 徒手绘图
an aid to 对……有所帮助，有助于
communication with one's own mind 和自己的大脑进行交流
stimulant ['stimjulənt] *n.* 刺激物，促进因素，激发
in earnest 认真地，真正地
have as 把……作为
alteration [ˌɔːltə'reiʃən] *n.* 变更，改造，改变
creative thought 创造性思维
reaction force 支反力，反作用力
buoyancy ['bɔiənsi] *n.* 浮力
centrifugal [sen'trifjugəl] *a.* 离心的
clutch bearing 离合器轴承
clutch facing 离合器面片，离合器摩擦片，离合器衬片
fluctuating stress 脉动应力，交变应力
thread [θred] *n.* 螺纹
box section 箱形截面，箱形断面
allowance [ə'lauəns] *n.* 余量，增加量
resonant vibration 共振
combination of properties 综合性能，性能组合
weldability 可焊性，焊接性
machinability [məʃiːnə'biliti] *n.* 机械加工性，切削加工性
noninterference [ˌnɔnintə'fiərəns] *n.* 不干涉；不(相互)干扰
locating of parts 零件的定位
tapped hole 螺纹孔

21　Material Selection

During recent years the selection of engineering materials has assumed great importance. Moreover, the process should be one of continual reevaluation. New materials often become available and there may be a decreasing availability of others. Concerns regarding environmental pollution, recycling and worker health and safety often impose new constraints. The desire for weight reduction or energy savings may dictate the use of different materials. Pressures from domestic and foreign competition, increased serviceability requirements, and customer feedback may all promote materials reevaluation. The extent of product liability actions, often the result of improper material use, has had a marked impact. In addition, the interdependence between materials and their processing has become better recognized. The development of new processes often forces reevaluation of the materials being processed. Therefore, it is imperative that design and manufacturing engineers exercise considerable care in selecting, specifying, and utilizing materials if they are to achieve satisfactory results at reasonable cost and still assure quality.

The first step in the manufacture of any product is design, which usually takes place in several distinct stages: (a) conceptual; (b) functional; (c) production design. During the conceptual-design stage, the designer is concerned primarily with the functions the product is to fulfill. Usually several concepts are visualized and considered, and a decision is made either that the idea is not practical or that the idea is sound and one or more of the conceptual designs should be developed further. Here, the only concern for materials is that materials exist that can provide the desired properties. If no such materials are available, consideration is given as to whether there is a reasonable prospect that new ones could be developed within cost and time limitations.

At the functional- or engineering-design stage, a practical, workable design is developed. Fairly complete drawings are made, and materials are selected and specified for the various components. Often a prototype or working model is made that can be tested to permit evaluation of the product as to function, reliability, appearance, serviceability, and so on. Although it is expected that such testing might show that some changes may have to be made in materials before the product is advanced to the production-design stage, this should not be taken as an excuse for not doing a thorough job of material selection. Appearance, cost, and reliability factors should be considered in detail, together with the functional factors. There is much merit to the practice of one very successful company which requires that all prototypes be built with the same materials that will be used in production and, insofar as possible, with the same manufacturing techniques. It is of little value to have a perfectly functioning prototype that cannot be manufactured economically in the expected sales volume, or one that is substantially different from what the production units will be in regard to quality and reliability. Also, it is much better for design engineers to do a complete job of material analysis, selection, and specification at the development stage of design rather than to leave it to the production-design stage, where changes may be made by others, possibly less knowledgeable about all of the functional aspects of the product.

At the production-design stage, the primary concern relative to materials should be that they are specified fully, that they are compatible with, and can be processed economically by, existing equipment, and that they are readily available in the needed quantities.

As manufacturing progresses, it is inevitable that situations will arise that may require modifications of the materials being used. Experience may reveal that substitution of cheaper materials can be made. In most cases, however, changes are much more costly to make after manufacturing is in progress than before it starts. Good selection during the production-design phase will eliminate the necessity for most of this type of change. The more common type of change that occurs after manufacturing starts is the result of the availability of new materials. These, of course, present possibilities for cost reduction and improved performance. However, new materials must be evaluated very carefully to make sure that all their characteristics are well established. One should always remember that it is indeed rare that as much is known about the properties and reliability of a new material as about those of an existing one. A large proportion of product failure and product-liability cases have resulted from new materials being substituted before their long-term properties were really known.

Product liability actions have made it imperative that designers and companies employ the very best procedures in selecting materials. The five most common faults in material selection have been: (a) failure to know and use the latest and best information available about the materials utilized; (b) failure to foresee, and take into account the reasonable uses for the product (where possible, the designer is further advised to foresee and account for misuse of the product, as there have been many product liability cases in recent years where the claimant, injured during misuse of the product, has sued the manufacturer and won); (c) the use of materials about which there was insufficient or uncertain data, particularly as to its long-term properties; (d) inadequate, and unverified, quality control procedures; and (e) material selection made by people who are completely unqualified to do so.

An examination of the faults above will lead one to conclude that there is no good reason why they should exist. Consideration of them provides guidance as to how they can be eliminated. While following the very best methods in material selection may not eliminate all product-liability claims, the use of proper procedures by designers and industries can greatly reduce their numbers.

From the previous discussion, it is apparent that those who select materials should have a broad, basic understanding of the nature and properties of materials and their processing.

Words and Expressions

reevaluation 重新评价
availability [əˌveiləˈbiliti] *n.* 可得到的,存在,可利用,使用价值
recycle [ˈriːsaikl] *v.* ; *n.* 再循环,回收,重复利用
dictate [dikˈteit] *v.* 规定,限定,确定,指挥
serviceability [ˌsəːvisəˈbiliti] *n.* 适用性,使用的可靠性,操作性能,可维修性
feedback [ˈfiːdbæk] *n.* 反馈
promote [prəˈməut] *v.* ; *n.* 促进,发扬,引起
liability [ˌlaiəˈbiliti] *n.* 责任,义务,赔偿责任,不利条件
product liability 产品责任(又称产品侵权损害赔偿责任,是指产品存在可能危及人身、财产安全的不合理危险,造成消费者人身或者除缺陷产品以外的其他财产损失后,缺陷产品的生产者、销售者应当承担的特殊的侵权法律责任)
marked impact 显著的影响
conceptual [kənˈseptjuəl] *a.* 概念上的
interdependence [ˌintədiˈpendəns] *n.* 互相依赖,相关性
imperative [imˈperətiv] *a.* ; *n.* 命令的,强制的,必不可少的
visualize [ˈvizjuəlaiz] *v.* 目测,检验,设想,具体化,直观化
prospect [ˈprɔspekt] *n.* 展望,预期,可能性,前景;*v.* 调查
substantially [səbˈstænʃəli] *ad.* 实质上,大体上,充分地
compatible [kəmˈpætəbl] *a.* 相容的,可共存的,协调的,相适应的
insofar as [insəuˈfɑːrəz] *conj.* 到这样的程度,在……情况下,既然,因为
inevitable [inˈevitəbl] *a.* 不可避免的,必然的,料得到的
reveal [riˈviːl] *v.* 展望,显示,揭示,揭露;*n.* 外露
claim [kleim] *n.* ; *v.* 要求,要求赔偿损失权,主张,断言
claimant [ˈkleimənt] *n.* 申请人,(根据权利)提出要求者,原告
sue [sjuː] *v.* 控诉,提出诉讼,提出请求

22 Selection of Materials

An ever-increasing variety of materials is now available, each having its own characteristics, applications, advantages, and limitations. The following are the general types of materials used in manufacturing today either individually or in combination.

1. Ferrous metals: carbon, alloy, stainless, tool, and die steels.
2. Nonferrous metals: aluminum, copper, nickel, titanium, and precious metals.
3. Plastics: thermoplastics, thermoset plastics, and elastomers.
4. Ceramics, glasses, graphite, and diamond.
5. Composite materials: reinforced plastics, metal-matrix and ceramic-matrix composites.
6. Nanomaterials, shape memory alloys, superconductors, and various other materials with unique properties.

As new materials are developed, the selection of appropriate materials becomes even more challenging. Aerospace structures, as well as products such as sporting goods, have been at the forefront of new material usage. The trend has been to use more titanium and composites for the airframe of commercial aircraft, with a gradual decline of the use of aluminum and steel. There are constantly shifting trends in the usage of materials in all products, driven principally by economic considerations as well as other considerations.

For instance, plastics are now widely used in numerous applications for such items as children's toys, automotive and electrical parts, and computer equipment, because of their durability and lower manufacturing costs. Ceramics have also become a hot item as they are used to make such things as computer chips, ceramic indexable inserts (see Fig. 40.3), and ceramic bearings. All these materials present the engineer and designer with a wide range of potential design and implementation challenges.

When selecting materials for products, we first consider their mechanical properties: strength, toughness, ductility, hardness, elasticity, and fatigue; then consider their physical properties: density, specific heat, thermal expansion and conductivity, melting point, and electrical and magnetic properties. Optimum designs often require a consideration of a combination of mechanical and physical properties. A typical example is the strength-to-weight and stiffness-to-weight ratios of materials for minimizing the weight of structural members. Aluminum, titanium, and reinforced plastics, for example, have higher ratios than steels and cast irons. Weight minimization is particularly important for aerospace and automotive applications, in order to improve performance and fuel economy.

Chemical properties also play a significant role in hostile (such as corrosive) as well as normal environments. Oxidation, corrosion, and flammability of materials are among the important factors to be considered.

Manufacturing properties of materials determine whether they can be cast, formed, machined, welded, and heat treated with relative ease. The method(s) used to process materials to the desired shapes can adversely affect the product's final properties, service life, and its cost.

Cost and availability of raw and processed materials and manufactured components are major concerns in manufacturing. If raw or processed materials or manufactured components are not available in the desired shapes, dimensions and quantities, substitutes and/or additional processing will be required, which can contribute significantly to product cost. For example, if we need a round bar of a certain diameter and it is not available in standard form, then we have to purchase a larger rod and reduce its diameter by some means, such as machining or grinding. It should be noted, however, that a product design can be modified to take advantage of standard dimensions of raw materials, thus avoiding additional manufacturing costs.

Reliability of supply, as well as demand, affects cost. Most countries import numerous raw materials that are essential for production. The United

States, for example, imports most of the cobalt, titanium, aluminum, nickel, natural rubber, and diamond that it needs.

Different costs are involved in processing materials by different methods. Some methods require expensive machinery, others require extensive labor, and still others require personnel with special skills, a high level of education, or specialized training.

The appearance of materials after they have been manufactured into products influences their appeal to the consumer. Color and surface texture are characteristics that we all consider when making a decision about purchasing a product.

Time- and service-dependent phenomena such as wear, fatigue, and dimensional stability are important. These phenomena can significantly affect a product's performance and, if not controlled, can lead to total failure of the product. Friction and wear, corrosion, and other phenomena can shorten a product's life or cause it to fail prematurely. Recycling or proper disposal of materials at the end of their useful service lives has become increasingly important in an age when we are more conscious of preserving resources and maintaining a clean and healthy environment. The proper treatment and disposal of toxic wastes and materials are also a crucial consideration.

Words and Expressions

ever-increasing variety of 不断增加的各种各样的,日益增加的各种
ferrous ['ferəs] **metal** 黑色金属,含铁金属
carbon, alloy, tool, and die steels 碳钢,合金钢,工具钢和模具钢
precious metals 贵金属(指金、银或铂等)
thermoplastics [,θə:mə'plæstiks] *n*. 热塑性塑料
thermoset plastics 热固性塑料
elastomer [i'læstəmə] *n*. 合成像胶,人造橡胶
reinforced plastics 增强塑料
graphite ['græfait] *n*. 石墨
ceramic matrix composite 陶瓷基复合材料

nanomaterial 纳米材料
shape memory alloy 形状记忆合金
sporting goods 体育用品
forefront ['fɔːfrʌnt] ***n.*** 最前部,(活动、趣味等的)中心
hot item 畅销商品
airframe ['ɛəfreim] ***n.*** 机身,航空器的构架
shifting trend 变化的趋势
computer chip 计算机芯片,电子芯片
ceramic indexable insert 陶瓷可转位刀片
flammability [ˌflæməˈbiləti] ***n.*** 易燃性,可燃性
formed 成形的,成形加工的
machined 经过机加工的(machine 的过去分词)
cobalt [kəˈbɔːlt] ***n.*** 钴(符号为 Co)
nickel [ˈnikl] ***n.*** 镍;***v.*** 镀镍于
availability [əˌveiləˈbiliti] ***n.*** 可用性
surface texture 表面结构(是表面粗糙度,表面波纹度,表面纹理和表面缺陷等的总称)
service-dependent 与使用有关的
prematurely [ˌpreməˈtjuəli] ***ad.*** 过早地
recycling [ˈriːˈsaikliŋ] ***n.*** 回收,再利用
disposal [disˈpəuzəl] ***n.*** 处理,处置
useful service lives 正常使用寿命
service life 使用寿命,使用期限,耐用年限
preserving resources 保护资源

23 Lathes

Lathes are machine tools designed primarily to do turning, facing, and boring. Very little turning is done on other types of machine tools, and none can do it with equal facility. Because lathes also can do drilling and reaming, their versatility permits several operations to be done with a single setup of the workpiece. Consequently, more lathes of various types are used in manufacturing than any other machine tool.

The essential components of a lathe are the bed, headstock, tailstock, carriage, leadscrew and feed rod, as shown in Fig. 23.1.

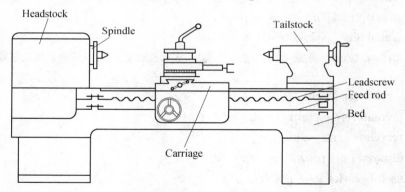

Figure 23.1 Schematic layout of a lathe

The bed is the backbone of a lathe. It is usually made of well-normalized or aged gray or nodular cast iron and provides a heavy, rigid frame on which all the other basic components are mounted. Two sets of parallel, longitudinal ways, inner and outer, are contained on the bed, usually on the upper side. Some makers use an inverted V-shape for all four ways, whereas others utilize one inverted V and one flat way in one or both sets. They are precision-machined to assure accuracy of alignment. On most modern lathes the ways are surface-hardened to resist wear and abrasion, but precaution should be taken in operating a lathe to assure that the ways are not damaged. Any inaccuracy in them usually means that the accuracy of the entire lathe is destroyed.

The headstock is mounted in a fixed position on the inner ways, usually at the left end of the bed. It provides a powered means of rotating the work at various speeds. Essentially, it consists of a hollow spindle, mounted in accurate bearings, and a set of transmission gears—similar to a truck transmission—through which the spindle can be rotated at a number of speeds. Most lathes provide from 8 to 18 speeds, usually in a geometric ratio. An increasing trend is to provide a continuously variable speed range through electrical or mechanical drives.

Because the accuracy of a lathe is greatly dependent on the spindle, it is of heavy construction and mounted in heavy bearings, usually preloaded tapered roller or ball types (see Fig. 12.1). The spindle has a hole extending through its length, through which long bar stock can be fed. The size of this hole is an important dimension of a lathe because it determines the maximum size of bar stock that can be machined when the material must be fed through spindle.

The chuck is connected to the spindle, where clamps and rotates the work. The three-jaw "self-centering" chuck (Fig. 23.2a) moves all of its jaws simultaneously to clamp or unclamp the work, and it is used for work with a round or hexagonal cross-section. The four "independent jaw" chuck (Fig. 23.2b) can clamp on the work by moving each jaw independent of the others. This chuck exerts a stronger hold on the work and it has the ability to center non-round shapes (squares, rectangles) exactly.

The tailstock consists, essentially, of three parts. A lower casting fits on the inner ways of the bed and can slide longitudinally thereon, with a means for clamping the entire assembly in any desired location. An upper casting fits on the lower one and can be moved transversely upon it, to permit aligning the tailstock and headstock spindles. The third major component of the tailstock is the tailstock quill. This is a hollow steel cylinder, usually about 51 to 76 mm (2 to 3 inches) in diameter, that can be moved several inches longitudinally in and out of the upper casting by means of a handwheel and screw.

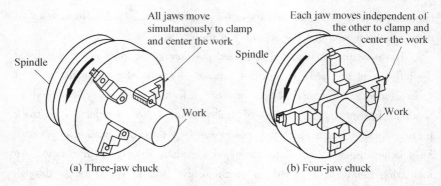

Figure 23.2 Chucks

The three-jaw and four-jaw chucks are normally suitable for short workpieces. The long workpieces are held between centers. The workpiece ends are provided with a center hole as shown in Fig. 23.3. Through these center holes the centers mounted in the spindle and the tailstock would locate the axis of the workpiece. However, these centers would not be able to transmit the motion to the workpiece from the spindle. For this purpose, generally a driver plate and a lathe dog would be used. In Fig. 23.3, the center located in the spindle is a dead center (Fig 23.4a) while that in the tailstock is a live center (Fig. 23.4b). The shank of the center is generally finished with a Morse taper which fits into the tapered hole of the spindle or tailstock.

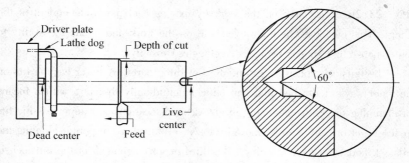

Figure 23.3 Workpiece being turned between centers in a lathe

The size of a lathe is designated by two dimensions. The first is known as the swing. It is approximately twice the distance between the line connecting the lathe centers and the nearest point on the ways. The second size dimension is the maximum distance between centers. The swing thus indicates the maximum workpiece diameter that can be turned in the lathe, while the distance between centers indicates the maximum length of workpiece that can be mounted between centers.

(a) Dead center   (b) Live center

Figure 23.4   Centers

Engine lathes are the type most frequently used in manufacturing. They are heavy-duty machine tools with all the components described previously and have power drive for all tool movements except on the compound rest. They commonly range in size from 305 to 610 mm (12 to 24 inches) swing and from 610 to 1 219 mm (24 to 48 inches) center distances, but swings up to 1 270 mm (50 inches) and center distances up to 3 658 mm (12 feet) are not uncommon. Most have chip pans and a built-in coolant circulating system. Smaller engine lathes—with swings usually not over 330 mm (13 inches)— also are available in bench type, designed for the bed to be mounted on a bench.

Although engine lathes are versatile and very useful, because of the time required for changing and setting tools and for making measurements on the workpiece, they are not suitable for quantity production. Often the actual chip-production time is less than 30% of the total cycle time.

Words and Expressions

turning ['tə:niŋ] *n.* 旋转,车削,车外圆

facing ['feisiŋ] *n.* 车平面,端面加工

boring ['bɔ:riŋ] *n*. 镗孔,镗削加工
reaming ['ri:miŋ] *n*. 铰孔
headstock ['hedstɔk] *n*. 主轴箱,动力箱
assembly [ə'sembli] *n*. 装配,组合,组件,部件
tailstock ['teilstɔk] *n*. 尾座,尾架
carriage ['kæridʒ] *n*. 溜板(在床身上使刀具作纵向移动的部件,一般由刀架、床鞍、溜板箱等组成)
leadscrew ['li:dskru:] *n*. 丝杠
feed rod 光杠
schematic layout 示意图,简图
way 导轨
abrasion [ə'breiʒən] *n*. 擦伤,刮去,磨损
geometric ratio 等比
bar stock 棒料
tapered ['teipəd] *a*. 锥形的,斜的
longitudinally [ˌlɔndʒi'tju:dinli] *ad*. 长度地,纵向地,轴向地
thereon [ðɛər'ɔn] *ad*. 在其中,在其上,关于那,紧接着
quill [kwil] *n*. 活动套筒,衬套,钻轴,空心轴
center ['sentə] *n*. 中心,顶尖;*v*. 定心,对中
center hole 中心孔,顶尖孔
axis ['æksis] *n*. 轴,轴线
driver plate 拨盘,有时也写为 drive plate
lathe dog 卡箍,鸡心夹头,有时也写为 dog
dead center 普通顶尖,死顶尖,固定顶尖
live center 活顶尖,活动顶尖,回转顶尖
morse taper 莫氏锥度
swing [swiŋ] *v*.;*n*. 摆动,最大回转直径
center distance 中心距,顶尖距
compound rest (车床)小刀架
bench type 台式

24 Drilling Operations

Drilling involves producing through or blind holes in a workpiece by forcing a tool, which rotates around its axis, against the workpiece. Consequently, the range of cutting from that axis of rotation is equal to the radius of the required hole.

Cutting Tools for Drilling Operations

In drilling operations, a cylindrical cutting tool, called a drill, is employed. The drill can have either one or more cutting edges and corresponding flutes, which can be straight or helical. The function of the flutes is to provide outlet passages for the chips generated during the drilling operation and also to allow lubricants and coolants to reach the cutting edges and the surface being machined. Following is a survey of the commonly used drills.

Twist Drills The twist drill is the most common type of drill. It has two cutting edges and two helical flutes that continue over the length of the drill body. The drill also consists of a neck and a shank that can be either straight or tapered.

Core Drills A core drill consists of the chamfer, body, neck, and shank, as shown in Fig. 24.1. A core drill has flat end. The chamfer can have three or four cutting edges. Core drills are employed for enlarging previously made holes and not for originating holes. This type of drill is characterized by greater productivity, high machining accuracy, and superior quality of the drilled surfaces.

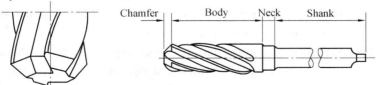

Figure 24.1　Core drill

Carbide Tipped Drills Most of the drills are made of high speed steel. However, for machining hard materials as well as for large volume production, carbide tipped drills are available. As shown in Fig. 24.2, the carbide tips of suitable geometry are clamped to the end of the tool to act as the cutting edges.

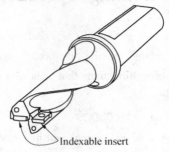

Figure 24.2 Drill with carbide inserts

Other Types of Drilling Operations

In addition to conventional drilling, there are other operations that are involved in the production of holes in the industrial practice. Following is a brief description of each of these operations.

Boring Boring involves enlarging a hole that has already been drilled. It is similar to internal turning and can, therefore, be performed on a lathe. There are also some specialized machine tools for carrying out boring operations. Those include the vertical boring machine, the jig boring machine, and the horizontal boring machine.

Counterboring Already existing holes in the parts can be further machined by counterboring as shown in Fig. 24.3a. In the counterboring operation, the hole is enlarged with a flat bottom to provide a space for the bolt head or a nut, so it would be entirely below the surface of the part.

Countersinking As shown in Fig. 24.3b, countersinking is done to enable accommodating the conical seat of a flathead screw so that the screw does not appear above the surface of the part.

Spot Facing Spot facing operation removes only a very small portion of material around the existing hole to provide a flat surface square to the hole axis, as shown in Fig.24.3c. This has to be done only in case where existing surface is not smooth and is usually performed on castings or forgings.

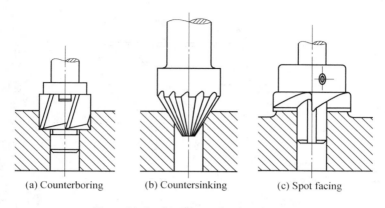

(a) Counterboring (b) Countersinking (c) Spot facing

Figure 24.3 Operations related to drilling

Reaming Reaming is an operation used to make an existing hole dimensionally more accurate than can be obtained by drilling alone. As a result of a reaming operation, a hole has a very smooth surface. The cutting tool used in this operation is known as the reamer.

Tapping Tapping is the process of cutting internal threads. The tool is called a tap.

Classification of Drilling Machines

Drilling operations can be carried out by using either electric hand drills or drilling machines. The latter differ in shape and size. Nevertheless, the tool always rotates around its axis while the workpiece is kept firmly fixed. This is contrary to the drilling operation on a lathe, where the workpiece is held in and rotates with the chuck. Following is a survey of the commonly used types of drilling machines.

Bench-type Drilling Machines Bench-type drilling machines are general-purpose, small machine tools that are usually placed on benches. This type of drilling machine includes an electric motor as the source of motion, which is transmitted via pulleys and belts to the spindle, where the tool is mounted (see Fig.24.4). The feed is manually generated by lowering a feed lever, which is designed to lower (or raise) the spindle. The workpiece is mounted on the machine table, although a special vise is sometimes used to hold the workpiece.

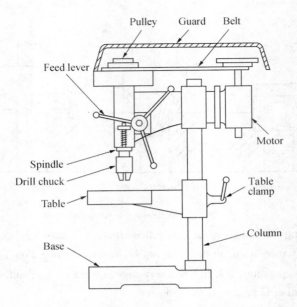

Figure 24.4　Sketch of a bench-type drilling machine

Upright Drilling Machines Depending upon the size, upright drilling machine tools can be used for light, medium, and even relatively heavy workpieces. It is basically similar to bench-type machines, the main difference being a longer cylindrical column fixed to the base. Along that column is an additional, sliding table for fixing the workpiece which can be locked in position at any desired height. The power required for this type is more than that for the bench-type drilling machines.

Radial Drills A radial drill is particularly suitable for drilling holes in large and heavy workpieces that are inconvenient to mount on the table of an upright drilling machine. A radial drilling machine has a main column, which is fixed to the base. The cantilever guide arm, which carries the drilling head, can be raised or lowered along the column and clamped at any desired position (see Fig. 24.5). The drilling head slides along the arm and provides rotary motion and axial feed motion. Again, the cantilever guide arm can be swung, thus enabling the tool to be moved in all directions according to a cylindrical coordinate system.

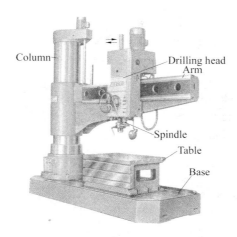

Figure 24.5　Radial drill

Words and Expressions

through or blind holes 通孔或盲孔
force against 把……压到……上,把……压入
symmetrical [si'metrikəl] *a*. 对称的,平衡的
flute [fluːt] *n*. 凹槽,(刀具的)排屑槽,容屑槽
outlet ['autlet] *n*. 排出口,流出口,排泄口,排水孔
coolant ['kuːlənt] *n*. 冷却液,散热剂,切削液,乳化液
twist drill 麻花钻头
shank [ʃæŋk] *n*. 刀柄,尾部,后部,杆,把手
high speed steel 高速钢
carbide ['kɑːbaid] *n*. 碳化物,硬质合金
carbide tip 硬质合金刀片,硬质合金片
carbide tipped drill 镶硬质合金钻头,硬质合金钻头
large volume production 大批量生产
indexable insert 可转位刀片
core drill 扩孔钻
chamfer ['tʃæmfə] *n*. 倒角,斜面,(扩孔钻的)切削部分
margin [mɑːˈdʒin] *n*. 刃带
guidance ['gaidəns] *n*. 引导,制导,向导,引导装置

enlarge [in'lɑːdʒ] *v.* 扩大,扩充,增大
originate [ə'ridʒineit] *v.* 起源,发源,引起,产生
tapping ['tæpiŋ] *n.* 攻螺纹
tap [tæp] *n.* 丝锥
vertical boring machine 立式镗床
jig boring machine 坐标镗床
horizontal boring machine 卧式镗床
counterbore [ˌkauntə'bɔː] *v.*,;*n.* (平底)锪孔,锪沉头孔
spot facing 锪端面
square to 与……成直角,垂直于
countersinking 锪锥孔
conical ['kɔnikəl] *a.* 圆锥形的
electric hand drill 手电钻
bench-type drilling machine 台式钻床
general purpose 通用的,多种用途的
feed lever 进给手柄
drill chuck 钻夹头,钻头夹盘
base 底座
upright ['ʌprait] *a.*;*ad.* 笔直的,竖立的,*n.* 支柱
upright drilling machine 立式钻床,也可以写为 vertical drilling machine
preset [priː'set] *v.*;*a.* 预调,预先设置
vise [vais] *n.* 虎钳,台钳;*v.* 钳住,夹紧
column ['kɔləm] *n.* 柱,柱状物,架,墩,(钻床等)立柱
medium-duty 中型的,中等的,中批生产
radial drill 摇臂钻床,也可写为 radial drilling machine
drilling head 钻床主轴箱,钻削动力头
cantilever ['kæntiliːvə] *n.* 悬臂;*v.* 使……伸出悬臂梁
cylindrical coordinate system 柱面坐标系统

25 Machining

Machining can be defined as the process of removal of the unwanted material (machining allowance) from the workpiece in the form of chips, so as to obtain a finished product of the desired size, shape, and surface quality. Machining includes turning, milling, grinding, drilling, etc.

Turning

Engine Lathes The engine lathe (see Fig. 23.1), one of the oldest metal removal machines, has a number of useful and highly desirable attributes. Today these lathes are used primarily in small shops where smaller quantities rather than large production runs are encountered.

Tolerances for the engine lathe depend primarily on the skill of the operator. The design engineer must be careful in using tolerances of an experimental part that has been produced on the engine lathe by a skilled operator. In redesigning an experimental part for production, economical tolerances should be used.

The engine lathe has been replaced in today's production shops by a wide variety of automatic lathes such as turret lathes, automatic screw machines, and automatic tracer lathes.

Turret Lathes Production machining equipment must be evaluated now, more than ever before, in terms of ability to repeat accurately and rapidly. Applying this criterion for establishing the production qualification of a specific method, the turret lathe merits a high rating.

In designing for low quantities such as 100 or 200 parts, it is most economical to use the turret lathe. In achieving the optimum tolerances possible on the turret lathe, the designer should strive for a minimum of operations.

Automatic Screw Machines Generally, automatic screw machines fall into two categories; single-spindle automatics and multiple-spindle automatics.

Originally designed for rapid, automatic production of screws and similar threaded parts, the automatic screw machine has long since exceeded the confines of this narrow field, and today plays a vital role in the mass production of a variety of precision parts.

Quantities play an important part in the economy of the parts machined on the automatic screw machine. Quantities less than 1 000 parts may be more economical to set up on the turret lathe than on the automatic screw machine. The cost of the parts machined can be reduced if the minimum economical lot size is calculated and the proper machine is selected for these quantities.

Automatic Tracer Lathes Since surface roughness depends greatly upon material turned, tooling, and feeds and speeds employed, minimum tolerances that can be held on automatic tracer lathes are not necessarily the most economical tolerances.

In some cases, tolerances of ± 0.05mm are held in continuous production using but one cut. Groove width can be held to ± 0.125mm on some parts. On high-production runs where maximum output is desirable, a minimum tolerance of ± 0.125mm is economical on both diameter and length of turn.

Milling

With the exceptions of turning and drilling, milling is undoubtedly the most widely used method of removing metal. Well suited and readily adapted to the economical production of any quantity of parts, the almost unlimited versatility of the milling process merits the attention and consideration of designers seriously concerned with the manufacture of their product.

As in any other process, parts that have to be milled should be designed with economical tolerances that can be achieved in production milling. If the part is designed with tolerances finer than necessary, additional operations will have to be added to achieve these tolerances—and this will increase the cost of the part.

Grinding

Grinding is one of the most widely used methods of finishing parts to extremely close tolerances and low surface roughness. There are five major types of grinding operations: surface (Fig. 25.1), cylindrical (Fig. 25.2), internal (Fig. 25.3), centerless, and abrasive belt grinding for precision work.

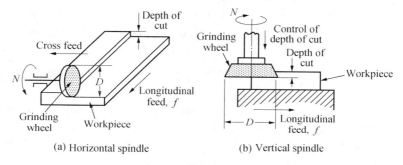

Figure 25.1 Surface grinding

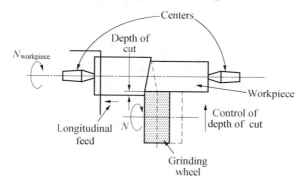

Figure 25.2 Cylindrical grinding

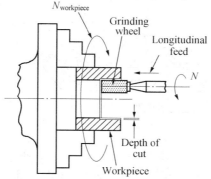

Figure 25.3 Internal grinding

Currently, there are grinders for almost every type of grinding operation. Particular design features of a part dictate to a large degree the type of grinding machine required. Where processing costs are excessive, parts redesigned to utilize a less expensive, higher output grinding method may be well worthwhile. For example, wherever possible the production economy of centerless grinding should be taken advantage of by proper design consideration.

Classes of grinding machines include the following: surface grinders (Fig. 25.4), cylindrical grinders (Fig. 25.5), internal grinders, centerless grinders, and tool and cutter grinders.

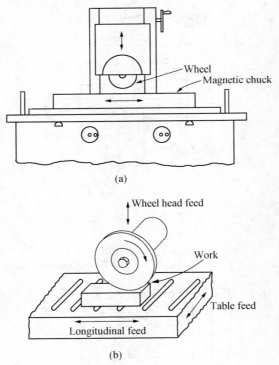

Figure 25.4 Sketch of a surface grinder

The surface grinders are for finishing all kinds of flat work, or work with plane surfaces which may be operated upon either by the periphery of the

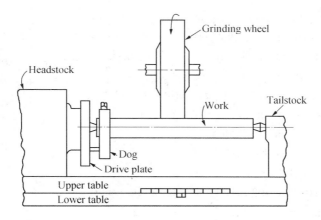

Figure 25.5 Sketch of a cylindrical grinder

grinding wheel or by the end face of the grinding wheel. These machines may have reciprocating or rotating tables. The cylindrical and centerless grinders are for straight cylindrical or taper work; thus shafts and similar parts are ground on cylindrical machines or centerless machines.

The internal grinders are used for grinding of precision holes, cylinder bores, and similar operations where bores of all kinds are to be finished.

Thread grinders are used for grinding precision threads for thread gages, and threads on precision parts where the concentricity between the diameter of the shaft and the pitch diameter of the thread must be held to close tolerances.

Words and Expressions

engine lathe 普通车床,卧式车床
removal [ri′mu:vəl] *n*. 除去,切削,切除
encounter [in′kauntə] *v*.; *n*. 遭遇,遇到,碰见
turret lathe 转塔车床
long since 很久以前,很久以来
automatic screw machine 自动螺纹车床,也称为自动车床(是通过凸轮来控制加工过程的自动加工机床。经过一定设置后可以长时间自动加工同一种产品。特别适合成批加工小零件,加工速度快,复杂零件一次加工成型,加工精度准确可靠,自动送料、生产效率高)
tracer [′treisə] *n*. 追踪装置,随动装置,仿形板

tolerance ['tɔlərəns] *n*. 公差,允许限度

merit ['merit] *n*. 优点,特征,价值;*v*. 值得,有……价值

confine [kən'fain] *v*. 限制在……范围内,封闭;*n*. 区域,范围

grinding ['graindiŋ] *n*.;*a*. 磨削(的)

cylindrical [si'lindrik(ə)l] *a*. 圆柱的,圆柱形的

centerless ['sentəlis] *a*. 无心的,没有心轴的

abrasive belt grinding 砂带磨削

longitudinal feed 纵向进给

cross feed 横向进给

depth of cut 切削深度,即背吃刀量(back engagement)

horizontal spindle 卧轴,卧式主轴,水平主轴

vertical spindle 立轴,立式主轴,垂直主轴

to a large degree 在很大程度上

magnetic chuck 电磁吸盘

wheel head 磨头(装有砂轮主轴并使其旋转的磨床部件)

dog 夹头,卡箍,鸡心夹

upper table 上工作台

lower table 下工作台

periphery [pə'rifəri] *n*. 外围,周边,边缘,圆周

reciprocating table 往复式工作台

taper ['teipə] *n*. 锥形,锥度

thread gage 螺纹量规

concentricity [ˌkɔnsən'trisəti] *n*. 同心,同心度

pitch [pitʃ] *n*. 间距,螺距,周节,螺旋线间隔

pitch diameter (螺纹)中径

bore [bɔː] *n*. 枪膛,汽缸筒,孔,孔径;*v*. 镗孔

26 Machine Tools

The general characteristics of the machine tools used in cutting operations are reviewed in this text. The equipment can be classified in different categories depending on their function and the accessories involved. The major categories are described below.

Lathes Lathes are generally considered to be the oldest machine tools and were developed in the 1750s. The basic operations carried out on a lathe are turning, boring, and facing.

The basic components of the most common lathe (*engine lathe*) are the bed, headstock, tailstock, and carriage (see Fig. 23.1). One end of the workpiece is clamped in a chuck (see Fig. 23.2) that is mounted on the spindle, the spindle rotates in the headstock. The other end of the workpiece is supported by the tailstock. The headstock contains the drive gears for the spindle speeds, and through suitable gearing and the feed rod, drives the carriage and cross slide assembly. The carriage provides motion parallel to the axis of rotation, and the cross slide provides motion normal to it. A cutting tool, attached to a tool post and to the carriage, removes material by traveling along the bed. The cutting operation is performed at a certain desired peripheral speed of the workpiece, feed rate, and depth of cut.

A further development is the *automatic lathe*. The movements and controls are actuated by various mechanical means, such as cams and numerical and computer control, thus reducing labor or requiring less skilled labor.

Turret lathes (developed in the 1850s) carry out multiple cutting operations, such as turning, boring, drilling, thread cutting, and facing. Various cutting tools, usually up to six are installed on a turret. The operation is quite versatile. It may be operated by hand or automatically and, once set up properly, it does not require skilled labor. There are a great variety of turret lathes available for different purposes.

Boring Machines Although boring operations can be carried out on a lathe, boring machines are used for large workpieces. These machines are either horizontal or vertical, and are available with a variety of features for boring and facing operations.

Jig borers are high-precision, vertical boring machines and are used for making jigs and fixtures, and for boring parts held in a jig.

Milling Machines These are among the most versatile and useful machine tools because they are capable of performing a variety of cutting operations. There is a large selection of milling machines (developed in the 1860s) available with numerous features. The basic types are outlined below.

Knee-and-column type milling machines are the most common and are used for general-purpose machining. The spindle, to which the milling cutter is attached, may be horizontal (for slab milling, as shown in Fig. 26.1), or vertical (for face milling, as shown in Fig. 26.2).

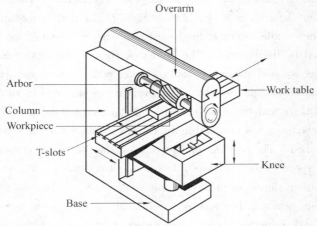

Figure 26.1 Horizontal knee and column type milling machine

Bed-type milling machines have the table mounted on a bed (replacing the knee) and can move only in the longitudinal direction. These machines are simple and rigidly constructed and are used for high-quantity production. The spindles may be horizontal or vertical and can be of duplex or triplex types (that is, with two or three spindles for simultaneous machining of two or three surfaces on a part). Other types of milling machines are also available, such as the planer type for heavy work and machines for special purposes.

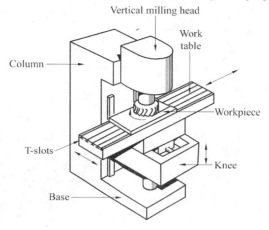

Figure 26.2 Vertical knee and column type milling machine

Among accessories for milling machines, one of the most widely used is the universal dividing head (Fig. 26.3). This is a fixture that rotates the workpiece to specified angles between individual machining steps. Typical uses are in milling parts with polygonal surfaces and in machining gear teeth.

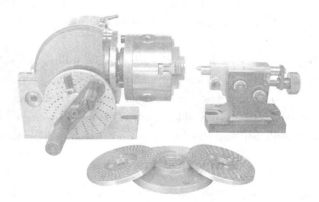

Figure 26.3 Universal dividing head

Planers and Shapers These machine tools are generally used for machining large flat surfaces although other shapes, such as grooves, can also be machined. In a planer, the workpiece travels under the cross-rail equipped with cutting tools, whereas in a shaper the tool travels and the workpiece is stationary (see Figs. 26.4 and 26.5).

Because of the size of the workpieces involved, planers are among the largest machine tools. Shapers may be horizontal or vertical and are also used for machining notches and keyways.

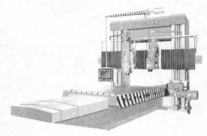 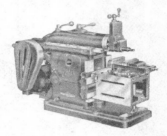

Figure 26.4　A planer　　　　　Figure 26.5　A shaper

Broaching Machines These machines push or pull broaches, either horizontally or vertically, to generate various external or internal surfaces. They are of relatively simple construction and are actuated hydraulically and have only linear motion of the broach. Because of the great variety of shapes and parts involved, broaching machines are manufactured in a wide range of designs and sizes.

Drilling Machines These machines are used for drilling holes, tapping, counterboring, reaming, and other general-purpose boring operations. They are usually vertical, the most common one being a drill press. Drilling machines are manufactured in a wide variety of sizes, from simple bench type units to radial drills and large production machines with multiple spindles.

Words and Expressions

accessory [æk'sesəri] *a*. 附属的,附带的,次要的; *n*. 附件,配件辅助装置
cross slide 横刀架,横拖板
attach M to N 把 M 连接[安装,固定,悬挂]到 N 上
tool post 刀架
peripheral [pə'rifərəl] *a*. 周围的,外围的; *n*. 外部设备,辅助设备

peripheral speed 圆周线速度
feed rate 进给速度
actuate ['æktjueit] *v.* 开动,驱动,使动作,操纵,驱使
means [mi:nz] *n.* 方法,方式,工具,设备,装置
jig [dʒig] *n.* 夹具,夹紧装置
jig borer 坐标镗床
jig and fixture 夹具
knee-and-column type milling machine 升降台式铣床
slab milling 平面铣削,周铣
arbor ['ɑ:bə] *n.* 刀杆,刀轴
overarm ['əuvɑ:m] *n.* 横梁
column (铣床)床身
face milling 端面铣削,面铣
work table 工作台
knee [ni:] *n.* (铣床的)升降台
vertical milling head 立铣头
dividing head 分度头
universal dividing head 万能分度头
duplex ['dju:pleks] *n.;a.* 双(的),双重(的),双联的,二部分的
triplex ['tripleks] *n.;a.* 三个部分(的),由三个组成(的),三联(的)
bed-type milling machine 工作台不升降式铣床,床身式铣床
planer type milling machine 龙门式铣床
polygonal [pɔ'ligənl] *a.* 多边形的,多角形的
planer ['pleinə] *n.* 龙门刨床
shaper ['ʃeipə] *n.* 牛头刨床
cross-rail ['krɔ:s'reil] *n.* 横导轨,横梁
broaching machine 拉床
broach [brəutʃ] *n.* 拉刀;*v.* 拉削
push broaching 推削
hydraulically [hai'drɔ:likəli] *ad.* 应用水力原理,液压地
tapping ['tæpiŋ] *n.* 攻丝,攻螺纹,车螺纹
drill press 钻床

27 Metal-Cutting Processes

Metal-cutting processes are extensively used in the manufacturing industry. They are characterized by the fact that the size of the original workpiece is sufficiently large that the final geometry can be circumscribed by it, and that the unwanted material is removed as chips, particles, and so on. The chips are a necessary means to obtain the desired tolerances, and surface roughness. The amount of scrap may vary from a few percent to 70% ~ 80% of the volume of the original work material.

Owing to the rather poor material utilization of the metal-cutting processes, the anticipated scarcity of materials and energy, and increasing costs, the development in the last decade has been directed toward an increasing application of metal-forming processes. However, die costs and the capital cost of machines remain rather high; consequently, metal-cutting processes are, in many cases, the most economical, in spite of the high material waste, which only has value as scrap. Therefore, it must be expected that the material removal processes will for the next few years maintain their important position in manufacturing. Furthermore, the development of automated production systems has progressed more rapidly for metal-cutting processes than for metal-forming processes.

Metal-cutting processes remove material from the surface of a workpiece by producing chips. This requires that the cutting tool material be harder than the workpiece material. The final geometry of the workpiece is thus determined from the geometry of the tool and the pattern of motions of the tool and the workpiece. The basic process is mechanical: actually, a shearing action combined with fracture.

As mentioned previously, the unwanted material in metal-cutting processes is removed by a rigid cutting tool, so that the desired geometry, tolerances, and surface roughness are obtained. Examples of processes in this group are turning, drilling, reaming, milling, shaping, planing, broaching, grinding, honing, and lapping.

Most of the cutting or machining processes are based on a two-dimensional surface creation, which means that two relative motions are necessary between the cutting tool and the work material. These motions are defined as the primary motion, which mainly determines the cutting speed, and the feed motion, which provides the cutting zone with new material.

In turning the primary motion is provided by the rotation of the workpiece, and in planing it is provided by the translation of the table; in turning the feed motion is a continuous translation of the tool, and in planing it is an intermittent translation of the tool. Figure 27.1 shows basic machining parameters in turning.

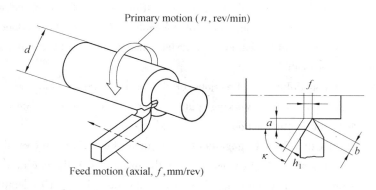

Figure 27.1 Basic machining parameters in turning

Cutting Speed The cutting speed v is the instantaneous velocity of the primary motion of the tool relative to the workpiece (at a selected point on the cutting edge).

The cutting speed for turning, drilling, and milling processes can be expressed as

$$v = \pi d n \ \mathrm{m/min} \tag{27.1}$$

where v is the cutting speed in m/min, d the diameter of the workpiece to be cut in meters, and n the workpiece or spindle rotation in rev/min. Thus v, d, and n may relate to the work material or the tool, depending on the specific kinematic pattern. In grinding the cutting speed is normally measured in m/s.

Feed The feed motion f is provided to the tool or the workpiece and, when added to the primary motion, leads to a repeated or continuous chip removal and the creation of the desired machined surface. The motion may proceed by steps or continuously. The feed speed v_f is defined as the instantaneous velocity of the feed motion relative to the workpiece (at a selected point on the cutting edge).

For turning and drilling, the feed f is measured per revolution (mm/rev) of the workpiece or the tool; for planing and shaping f is measured per stroke (mm/stroke) of the tool or the workpiece. In milling the feed is measured per tooth of the cutter f_z (mm/tooth); that is, f_z is the displacement of the workpiece between the cutting action of two successive teeth. The feed speed v_f (mm/min) of the table is therefore the product of the number of teeth z of the cutter, the revolutions per minute of the cutter n, and the feed per tooth ($v_f = nzf_z$).

A plane containing the directions of the primary motion and the feed motion is defined as the working plane, since it contains the motions responsible for the cutting action.

Depth of Cut (Back Engagement) In turning (Fig. 27.1) the depth of cut a (sometimes also called engagement) is the distance that the cutting edge engages or projects below the original surface of the workpiece. The depth of cut determines the final dimensions of the workpiece. In turning, with an axial feed, the depth of cut is a direct measure of the decrease in radius of the workpiece and with radial feed the depth of cut is equal to the decrease in the length of workpiece.

Chip Thickness The chip thickness h_1 in the undeformed state is the thickness of the chip measured perpendicular to the cutting edge and in a plane perpendicular to the direction of cutting (see Fig. 27.1). The chip thickness after cutting (i.e., the actual chip thickness h_2) is larger than the undeformed chip thickness, which means that the cutting ratio or chip thickness ratio $r = h_1/h_2$ is always less than unity.

Chip Width The chip width b in the undeformed state is the width of the chip measured along the cutting edge in a plane perpendicular to the direction of cutting.

Area of Cut For single-point tool operations, the area of cut A is the product of the undeformed chip thickness h_1 and the chip width b (i.e., $A = h_1 b$). The area of cut can also be expressed by the feed f and the depth of cut a as follows:

$$h_1 = f \sin\kappa \quad \text{and} \quad b = a/\sin\kappa \qquad (27.2)$$

where κ is the tool cutting edge angle.

Consequently, the area of cut is given by

$$A = fa \qquad (27.3)$$

Words and Expressions

workpiece ['wəːkpiːs] *n.* 工件
circumscribe ['səːkəmskraib] *v.* 与……外接,确定……的范围,限制
chip [tʃip] *n.* 切屑; *v.* 切,削,刨
scrap [skræp] *n.* 碎片,切屑,废品,废渣; *a.* 碎片的; *v.* 废弃,使成碎屑
work material 工件材料
scarcity ['skɛəsiti] *n.* 缺乏,不足,稀少,供不应求
die [dai] *n.* 模具,冲模,压模
primary motion 主运动
feed motion 进给运动
intermittent [ˌintəˈmitənt] *a.* 间歇的,间断的,断续的
instantaneous [ˌinstənˈteinjəs] *a.* 瞬时作用的,同时发生的
stroke [strəuk] *n.* 打击,冲程,行程,循环,周期
feed per tooth 每齿进给量
engagement [inˈgeidʒmənt] *n.* 啮合,吃刀量
back engagement 背吃刀量
engage [inˈgeidʒ] *v.* 啮合,切入
undeformed [ˈʌndiˈfɔːmd] *a.* 无变形的,未变形的
unity [juːniti] *n.* 一,单一,唯一,统一,单位
tool cutting edge angle 主偏角

28 Milling Operations

After lathes, milling machines are the most widely used for manufacturing applications. In milling, the workpiece is fed into a rotating milling cutter, which is a multi-edge tool as shown in Fig. 28.1. Metal removal is achieved through combining the rotary motion of the milling cutter and linear motions of the workpiece simultaneously.

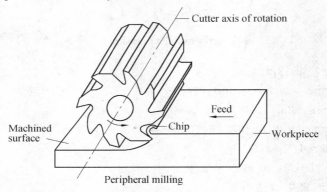

Figure 28.1 Schematic diagram of a milling operation

Each of the cutting edges of a milling cutter acts as an individual single-point cutter when it engages with the workpiece metal. Therefore, each of those cutting edges has appropriate rake and relief angles. Since only a few of the cutting edges are engaged with the workpiece at a time, heavy cuts can be taken without adversely affecting the tool life. In fact, the permissible cutting speeds and feeds for milling are three to four times higher than those for turning or drilling. Moreover, the quality of surfaces machined by milling is generally superior to the quality of surfaces machined by turning, shaping, or drilling.

As far as the directions of cutter rotation and workpiece feed are concerned, milling is performed by either of the following two methods.

(1) **Up milling** (**conventional milling**). In up milling the cutting tool

rotates in the opposite direction to the table movement, the chip starts as zero thickness and gradually increases to the maximum size as shown in Fig. 28.2a. This tends to lift the workpiece from the table. There is a possibility that the cutting tool will rub the workpiece before starting the removal. However, the machining process involves no impact loading, thus ensuring smoother operation of the machine tool.

The initial rubbing of the cutting edge during the start of the cut in up milling tends to dull the cutting edge and consequently have a lower tool life. Also since the cutter tends to cut and slide alternatively, the quality of machined surface obtained by this method is not very high. Nevertheless, up milling is commonly used in industry, especially for rough cuts.

(2) **Down milling (climb milling)**. In down milling the cutting tool rotates in the same direction as that of the table movement, and the chip starts as maximum thickness and goes to zero thickness gradually as shown in Fig. 28.2b.

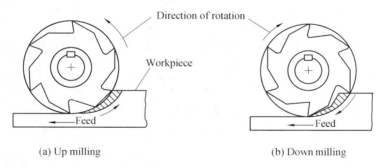

(a) Up milling　　　　　　　　(b) Down milling

Figure 28.2　Milling methods

The advantages of this method include higher quality of the machined surface and easier clamping of workpieces, since cutting forces act downward. Down milling also allows greater feeds per tooth and longer tool life between regrinds than up milling. But, it cannot be used for machining castings (see Fig. 28.3) or hot rolled steel, since the hard outer scale will damage the cutter.

There are a large variety of milling cutters to suit specific requirements. The cutters most generally used, shown in Fig. 28.4, are classified according to their general shape or the type of work they will do.

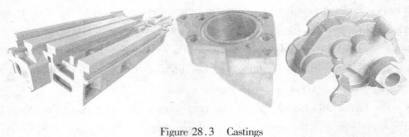

Figure 28.3 Castings

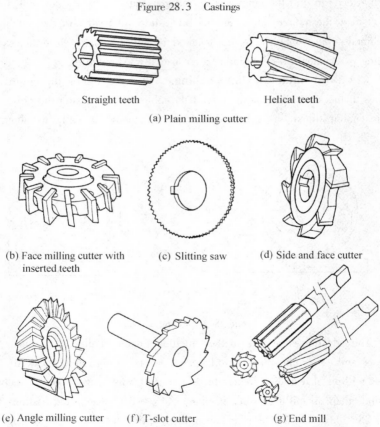

Figure 28.4 Types of milling cutters

Plain milling cutters. They are basically cylindrical with the cutting teeth on the periphery, and the teeth may be straight or helical, as shown in Fig. 28.4a. The helical teeth generally are preferred over straight teeth because the tooth is partially engaged with the workpiece as it rotates. Consequently, the cutting force variation will be smaller, resulting in a smoother operation. The cutters are generally used for machining flat surfaces.

Face milling cutters. They have cutting edges on the face and periphery. The cutting teeth, such as carbide inserts, are mounted on the cutter body as shown in Fig. 28.4b.

Most larger-sized milling cutters are of inserted-tooth type. The cutter body is made of ordinary steel, with the teeth made of high speed steel, cemented carbide, or ceramics, fastened to the cutter body by various methods. Most commonly, the teeth are indexable carbide or ceramic inserts.

Slitting saws. They are very similar to a saw blade in appearance as well as function (Fig. 28.4c). The thickness of these cutters is generally very small. The cutters are employed for cutting off operations and deep slots.

Side and face cutters. They have cutting edges not only on the periphery like the plain milling cutters, but also on both the sides (Fig. 28.4d). As was the case with the plain milling cutter, the cutting teeth can be straight or helical.

Angle milling cutters. They are used in cutting dovetail grooves and the like. Figure 28.4e indicates a milling cutter of this type.

T-slot cutters. T-slot cutters (Fig. 28.4f) are used for milling T-slots such as those in the milling machine table (see also Figs. 26.1 and 26.2).

End mills. There are a large variety of end mills. One of distinctions is based on the method of holding, i.e., the end mill shank can be straight or tapered. The straight shank is used on end mills of small size. The tapered shank is used for large cutter sizes. The cutter usually rotates on an axis perpendicular to the workpiece surface.

Figure 28.4g shows three kinds of end mills. The cutter can remove material on both its end and its cylindrical cutting edges. Vertical milling machines and machining centers (see Fig. 28.5) can be used for end milling workpieces of various sizes and shapes. The machines can be programmed such that the cutter can follow a complex set of paths that optimize the whole machining operation for productivity and minimum cost.

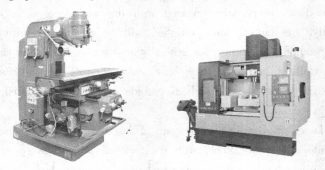

Figure 28.5　(*Left*) Vertical milling machine. (*Right*) Vertical machining center

Words and Expressions

milling cutter 铣刀

multi-edge tool 多刃刀具(有多个主切削刃参加切削的刀具)也可以写为 multi-point tool

machined surface 已加工表面

peripheral milling 圆周铣削, 周铣

schematic [ski'mætik] *a*. 示意性的, 图解的, 图表的

schematic diagram 原理图, 示意图

single-point cutter 单刃刀具(切削时只用一个主切削刃的刀具)

cutting edge 切削刃, 刀刃

rake angle 前角

relief angle 后角

chip [tʃip] *n*. 切屑

heavy cut 重切削, 强力切削

up milling 逆铣

conventional milling 逆铣

down milling 顺铣

climb milling 顺铣

table 工作台

table feed 工作台进给
cutting life 刀具寿命
regrind [ri'graind] *v.* 重新研磨，再次磨削
casting ['kɑːstiŋ] *n.* 铸件，铸造
hot rolled steel 热轧钢
outer scale 外层氧化皮
helical teeth 螺旋齿
plain milling cutter 圆柱铣刀
face milling cutter 端铣刀，面铣刀
face milling cutter with inserted teeth 镶齿端铣刀，镶齿面铣刀
slitting saw 锯片铣刀
cut off 切断
side and face cutter 三面刃铣刀
angle milling cutter 角度铣刀
T-slot cutter T形槽铣刀
end mill 立铣刀
cutter body 刀体
cemented carbide 硬质合金
fasten ['fɑːsn] *v.* 固定，使坚固或稳固
saw blade 锯条
carbide insert 硬质合金刀片
indexable ceramic insert 陶瓷可转位刀片
as was the case with 正如，就像
dovetail groove 燕尾槽
and the like 等等
see also 参见，另请参阅
tapered shank 锥柄
vertical milling machine 立式铣床
machining center 加工中心

29 Gear Manufacturing Methods

Gear Planing The shape of the space between gear teeth is complex and varies with the number of teeth on the gear as well as tooth module, so most gear manufacturing methods generate the tooth flank instead of forming.

Gear planing uses a reciprocating rack, stroking in the direction of the helix on a gear with a gradual generation of form as the rack effectively rolls round the gear blank. The rack is relieved out of contact for the return stroke as in normal shaping or planing. It has the great advantage that the cutting tool is a simple rack with (nearly) straight sided teeth which can easily be ground accurately. This method is little used for high volume production because it is relatively slow in operation; for jobbing purposes the slow stroking rate does not matter and low tool costs give an advantage where unusual sizes or profile modifications are required.

Gear Shaping Gear shaping is inherently similar to gear planing but uses a circular cutter instead of a rack (Fig. 29.1) and the resulting reduction in the reciprocating inertia allows much higher stroking speeds; modern gear shapers cutting car gears can run at 2000 cutting strokes per minute.

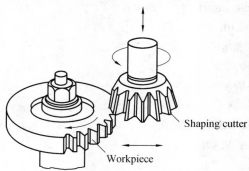

Figure 29.1 Gear shaping process

In gear shaping process, the tool and workpiece move tangentially typically 0.5 mm for each stroke of the cutter. On the return stroke the cutter must be retracted about 1 mm to give clearance otherwise tool rub occurs on the backstroke and failure is rapid.

The advantages of gear shaping are that production rates are relatively high and that it is possible to cut right up to a shoulder. Unfortunately, for helical gears (see Fig. 30.1b), a helical guide is required to impose a rotational motion on the stroking motion (as shown in Fig. 29.2); such helical guides cannot be produced easily or cheaply so the method is only suitable for long runs with helical gears since special cutters and guides must be manufactured for each different helix angle. A great advantage of gear shaping is its ability to cut internal gears such as those required for large epicyclic drives.

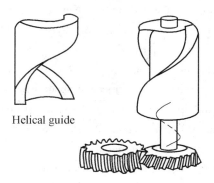

Figure 29.2   Helical gear shaping process

Gear Hobbing   Gear hobbing, the most used metal cutting method, uses the rack generating principle but avoids slow reciprocation by mounting many "racks" on a rotating cutter. The "racks" are displaced axially to form a gashed worm (Fig. 29.3). The gear hobbing process is shown in Fig. 29.4.

Figure 29.3   Gear hob

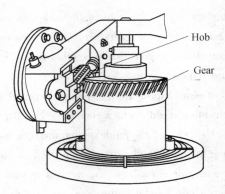

Figure 29.4 Gear hobbing process

Metal removal rates are high since no reciprocation of hob or workpiece is required and so cutting speeds of 40 m/min can be used for conventional hobs and up to 150 m/min for carbide hobs. Typically with a 100 mm diameter hob the rotation speed will be 100 rpm and so a twenty tooth workpiece will rotate at 5 rpm. Each revolution of the workpiece will correspond to 0.75 mm feed so the hob will advance through the workpiece at about 4 mm per minute. For car production roughing multiple start hobs can be used with coarse feeds of 3 mm per revolution so that 100 rpm on the cutter, a two-start hob and a 20 tooth gear will give a feed rate of 30 mm/minute.

Gear Broaching Gear broaching is not usually used for helical gears but is useful for internal spur gears; the principle use of broaching in this context is for internal splines which cannot easily be made by any other method. As with all broaching the method is only economic for large quantities since tooling costs are high.

Gear broaching gives high accuracy and low surface roughness but like all cutting processes is limited to "soft" materials which must be subsequently case-hardened or heat treated, giving distortion.

Gear Shaving Gear shaving is used as finishing processes for gears in the "soft" state. The objective is to improve surface roughness and profile by mating the roughed-out gear with a "cutter" which will improve form.

A gear shaving cutter looks like a gear which has extra clearance at the root (for swarf and coolant removal) and whose tooth flanks have been grooved to give cutting edges. It is run in mesh with the rough gear with crossed axes so that there is in theory point contact with a relative velocity along the teeth giving scraping action, as indicated in Fig. 29.5. The gear shaving cutter teeth are relatively flexible in bending and so will only operate effectively when they are in double contact between two gear teeth. Cycle times can be less than half a minute and the machines are not expensive but cutters are delicate and difficult to manufacture.

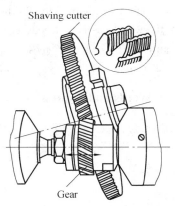

Figure 29.5 Gear shaving process

Gear Grinding Gear grinding is extremely important because it is the main way hardened gears are machined. When high accuracy is required it is not sufficient to pre-correct for heat treatment distortion and grinding is then necessary.

The simplest approach to gear grinding is form grinding. The wheel profile is dressed accurately to shape using single point diamonds which are controlled by templates cut to the exact shape required. The profiled wheel is then reciprocated axially along the gear, when one tooth shape has been finished, involving typically 100 micron metal removal, the gear is indexed to the next tooth space. This method is fairly slow but gives high accuracy consistently. Setting up is lengthy because different dressing templates are needed if module, number of teeth, helix angle, or profile correction are changed.

The fastest gear grinding method uses the same principle as gear hobbing but replaces the hob by a grinding wheel which is a rack in section. Only single start worms are cut on the wheel but gear rotation speeds are high, 100 rpm typically, so it is difficult to design the drive system to give accuracy and rigidity. Accuracy of the process is reasonably high although there is a tendency for wheel and workpiece to deflect variably during grinding so the wheel form may require compensation for machine deflection effects. Generation of a worm shape on the grinding wheel is a slow process since a dressing diamond must not only form the rack profile but has to move axially as the wheel rotates. Once the wheel has been trued, gears can be ground rapidly until redressing is required. This is the most popular method for high production rates with small gears.

Words and Expressions

flank [flæŋk] *n*. 侧面,后面,边,外侧
tooth flank 齿面
planing ['pleiniŋ] *n*.; *a*. 刨削(的)
rack [ræk] *n*. 齿条
helix ['hi:liks] *n*.; *a*. 螺旋,螺旋线
helical gear 斜齿轮
gradual ['grædjuəl] *a*. 逐渐的,渐渐的,顺序变化的
generation [,dʒenə'reiʃən] *n*. 生产,发生,制造,展成
jobbing ['dʒɔbiŋ] *n*. 做临时工,重复性很小的工作
jobbing shop 修理车间
shaping ['ʃeipiŋ] *v*. 刨削,压力加工; *a*. 成形的
gear shaping 插齿
involute ['invəlu:t] *n*. 渐开线
retract [ri'trækt] *v*. 缩回,缩进,收缩,取消
backstroke ['bækstrəuk] *n*.; *v*. 返回行程,回程
internal gear 内齿轮,内齿圈
epicyclic ['episaiklik] *a*. 周转圆的,外摆线的
epicyclic gear 行星齿轮

hobbing [ˈhɔbiŋ] *n*. 滚削
gear hobbing 滚齿
gash [gæʃ] *n*. 裂口, 裂纹; *v*. 划开, 造成深长切口
hob [hɔb] *n*. 滚刀
spur [spə:] *n*. 齿, 正齿, 支承物
spur gear 直齿轮
in this context 由于这个原因, 在这个意义上, 在这方面, 在这种情况下
spline [splain] *v*. 把……刻出键槽, 用花键连接; *n*. 花键, 键槽, 样条函数
case-harden 表面淬火, 表面渗碳硬化
rough [rʌf] *a*.; *ad*. 粗糙(的), 不光的, 粗(未)加工的; *v*. 粗制, 粗加工
shaving [ˈʃeiviŋ] *n*. 剃齿, 剃齿法
roughed out 经过粗加工的
swarf [swɔ:f] *n*. 细铁屑
in mesh 齿轮互相啮合
scraping [ˈskreipiŋ] *n*. 刮削
micron [ˈmaikrɔn] *n*. 微米, 也可写为 micrometer
form grinding 成形磨削
dress [dres] *v*. 修整, 精整, 打磨
dressing [ˈdresiŋ] *n*. 砂轮的修锐(产生锋锐的磨削面), 对于常用的砂轮来说, truing 和 dressing 通常在同一过程中完成, 这二者的综合作用也被称为 dressing
worm [wə:m] *n*. 蜗杆; *v*. 蠕动, 爬行
dresser 修整器(为了获得一定的几何形状和锐利磨削刃, 用金刚石等工具对砂轮表面进行整修的装置)
truing [ˈtru:iŋ] *n*. 砂轮的整形(产生确定的几何形状)
true [tru:] *a*. 真正的; *v*. 调整, 精密修整, 砂轮的整形(产生确定的几何形状)
dressing diamond 金刚石修整工具(通过钎焊或压入的方法镶嵌到钢制载体上的金刚石)

30 Gear Materials

Gears (see Fig. 30.1) are manufactured from a wide variety of materials, both metallic as well as nonmetallic. As is the case with all materials used in design, the material chosen for a particular gear should be the cheapest available that will ensure satisfactory performance. Before a choice is made, the designer must decide which of several criteria is most important to the problem at hand. If high strength is the prime consideration, a steel should usually be chosen rather than cast iron. If wear resistance is the most important consideration, a nonferrous material is preferable to a ferrous one. As still another example of how a choice can be made, for problems involving noise reduction, nonmetallic materials perform better than metallic ones. However, as is true in most design problems, the final choice of a material is usually a compromise. In other words, the material chosen will conform reasonably well to all the requirements mentioned previously, although it will not necessarily be the best in any one area. To conclude this discussion we will consider the characteristics of various metallic and nonmetallic gear materials according to their general classifications.

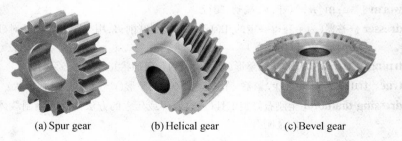

(a) Spur gear (b) Helical gear (c) Bevel gear

Figure 30.1 Gears

Cast Irons Cast iron is one of the most commonly used gear materials. Its low cost, ease of casting, good machinability, high wear resistance, and good noise abatement property make it a logical choice. The primary disadvantage of cast iron as a gear material is its low tensile strength, which makes the gear tooth weak in bending and necessitates rather large teeth.

Another type of cast iron is nodular iron, which is made of cast iron to which a material such as magnesium or cerium has been added. The result of this alloying is a material having a much higher tensile strength while retaining the good wear and machining characteristics of ordinary cast iron.

Very often the combination of cast iron gear and a steel pinion will give a well balanced design with regard to cost, strength, and wear.

Steels Steel gears are usually made of plain carbon steels or alloy steels. They have the advantage, over cast iron, of higher strength without undue increase in cost. However, they usually require heat treatment to produce a surface hard enough to give satisfactory resistance to wear. Unfortunately, the heat treatment process usually produces distortion of the gear, with the result that the gear load is not uniformly distributed across the gear tooth face. Since alloy steels are subject to less distortion due to heat treatment than carbon steels, they are often chosen in preference to the carbon steels.

Gears are often through-hardened by water or oil quenching in order to increase their resistance to wear. Tempering is usually performed immediately after quenching and involves reheating the gear to a temperature of 200℃ to 700℃ and then slowly cooling it in air back to room temperature. If a low degree of hardness is satisfactory, through-hardening is probably the most desirable heat treatment process to be used because of its inexpensiveness.

In many cases, the bulk of the part requires only moderate strength although the surface must have a very high hardness. In gear teeth, for example, high surface hardness is necessary to resist wear as the mating teeth come into contact several million times during the expected life of the gears. At each contact, a high stress develops at the surface of teeth. For applications such as this, case hardening is used, the surface (or case) of the part is given a high hardness to a depth of perhaps $0.25 \sim 1.00$ mm, although the interior of the part (the core) is affected only slightly, if at all. The advantage of surface hardening is that as the surface receives the required wear-resisting hardness, the core of the part remains in a more ductile form, resistant to impact and fatigue. The processes used most often for case hardening are flame hardening, induction hardening, carburizing, nitriding, cyaniding, and carbo-nitriding.

Figure 30.2 shows a drawing of a typical case-hardened gear-tooth section, clearly showing the hard case surrounding the softer, more ductile core.

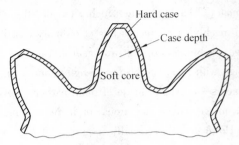

Figure 30.2 Typical case-hardened gear-tooth section

Nonferrous Metals Copper, zinc, aluminum, and titanium are materials used to obtain alloys that are useful gear materials. The copper alloys, known as bronzes, are perhaps the most widely used. They are useful in situations where corrosion resistance is important and also where large sliding velocities exist. Because of their ability to reduce friction and wear, they are usually used as the material for making the worm wheel in a worm gear speed reducer, which is shown in Fig. 30.3.

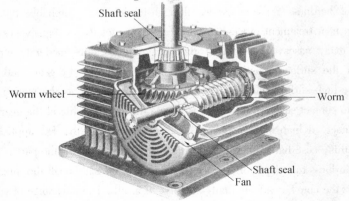

Figure 30.3 Worm gear speed reducer

Nonmetallic Materials Gears have been manufactured of nonmetallic materials for many years. Nylon, various types of plastics, and so on, have been used. The advantages obtained by using these materials are quiet operation, internal lubrication, dampening of shock and vibration, and manufacturing economy. Their primary disadvantages are lower load carrying capacity and low heat conductivity, which results in heat distortion of the teeth and may result in a serious weakening of the gear teeth.

Recently thermoplastic resins, with glass-fiber reinforcement and a lubricant as additives, have been used as gear materials. The composite material has resulted in greater load carrying capacity, a reduced thermal expansion, greater wear resistance and fatigue endurance.

Words and Expressions

nonmetallic ['nɔnmi'tælik] *a.* 非金属的; *n.* 非金属物质
metallic [mi'tælik] *a.* 金属的,金属制的
as is the case 通常是这样的
criteria [krai'tiəriə] *n.* criterion 的复数
bevel gear 锥齿轮(分度曲面为圆锥面的齿轮)
cast iron 铸铁
nonferrous ['nɔn'ferəs] *a.* 非铁的
nonferrous metal 有色金属
ferrous ['ferəs] *a.* 铁的,铁类的
ferrous material 钢铁材料,黑色金属
classification [ˌklæsifi'keiʃən] *n.* 分类,归类,类别
abatement [ə'beitmənt] *n.* 减少,减轻,降低,抑制,削弱
necessitate [ni'sesiteit] *v.* 需要,使成为必要,以……为条件,迫使
nodular ['nɔdjulə] *a.* 节状的,球状的,团状的
nodular cast iron 球墨铸铁
magnesium [mæg'ni:ziəm] *n.* 镁
cerium ['siəriəm] *n.* 铈
pinion ['pinjən] *n.* 小齿轮,传动齿轮
well balanced 各方面协调的,匀称的,平衡的

carbon steel 碳素钢，碳钢
undue [ʌn'dju] *a*. 过度的，不相称的，不适当的，未到期的，不正当的
in preference to M 优先于 M，(宁取……)而不取 M，比 M 好
through-hardening *n*. 整体淬火，淬透
inexpensiveness ['iniks'pensivnis] *n*. 花费不多，廉价，便宜
bulk of a part 零件的主体部分
moderate strength 中等强度
hard ease 表面硬化层
case depth 硬化层深度，表面硬化深度
carburize ['kɑːbjuraiz] *v*. 渗碳，碳化
cyanide ['saiənaid] *n*. 氰化物；用氰化物处理
nitride ['naitraid] *n*.；*v*. 氮化，渗氮
carbo-nitriding 碳氮共渗
flame hardening 火焰淬火
induction [in'dʌkʃən] *n*. 引导，感应，电感，归纳
induction hardening 高频淬火，感应淬火
zinc [ziŋk] *n*. 锌
worm 蜗杆
worm wheel 蜗轮
worm gear 蜗轮，蜗轮传动装置
worm gear speed reducer 蜗轮减速器
diecasting 压铸件，压铸法
nylon ['nailən] *n*. 尼龙，尼龙织品
dampen ['dæmpən] *v*. = **damp** 阻尼，减振，缓冲，抑制，衰减
heat conductivity 导热性，导热率
thermoplastic ['θəːməu'plæstik] *n*.；*a*. 热塑性，塑性，热塑性的
resin ['rezin] *n*. 树脂，树脂制品；*v*. 用树脂处理
glass-fiber reinforcement 玻璃纤维增强
endurance [in'djuərəns] *n*. 忍耐，持久(性)，耐用度，耐疲劳强度，寿命

31　Dimensional Tolerances and Surface Roughness

　　Today's technology requires that parts be specified with increasingly exact dimensions. Many parts made by different companies at widely separated locations must be interchangeable, which requires precise size specifications and production.

　　The technique of dimensioning parts within a required range of variation to ensure interchangeability is called tolerancing. Each dimension is allowed a certain degree of variation within a specified zone, or tolerance. For example, a part's dimension might be expressed as 20 ± 0.05, which allows a tolerance (variation in size) of 0.10 mm.

　　Tolerance is an undesirable, but permissible, deviation from a desired, theoretical dimension in recognition that no part can be made exactly to a specified dimension, except by chance, and such is neither necessary nor economical.

　　A tolerance should be as large as possible without interfering with the function of the part to minimize production costs. Manufacturing costs increase as tolerances become smaller.

　　There are three methods of specifying tolerances on dimensions: Unilateral, bilateral, and limit forms. When plus-or-minus tolerancing is used, it is applied to a theoretical dimension called the basic dimension. When dimensions can vary in only one direction from the basic dimension (either larger or smaller) tolerancing is unilateral. Tolerancing that permits variation in both directions from the basic dimension (larger and smaller) is bilateral. Tolerances may also be given in limit form, with dimensions representing the largest and smallest sizes for a feature.

　　Some tolerancing terminology and definitions are given below.

　　Tolerance: the difference between the limits prescribed for a single feature.

Basic size: the theoretical size, from which limits or deviations are calculated.

Deviation: the difference between the hole or shaft size and the basic size.

Upper deviation: the difference between the maximum permissible size of a part and its basic size.

Lower deviation: the difference between the minimum permissible size of a part and its basic size.

Actual size: the measured size of the finished part.

Fit: the tightness between two assembled parts. The three types of fit are: clearance, interference and transition.

Without needing to know how to operate a particular machine to attain the desired degree of surface roughness, there are certain aspects of all these methods which should be understood by the design engineer. Knowledge of such facts as degree of roughness obtained by any operation, and the economics of attaining a smoother surface with each operation, will aid him in deciding just which surface roughness to specify.

Because of its simplicity, the arithmetical average *Ra* has been adopted internationally and is widely used. The applications of surface roughness *Ra* are described in the following paragraphs.

0.2μm The surface roughness is used for the surface of hydraulic struts, for hydraulic cylinders (see Fig. 31.1), pistons and piston rods for O-ring packings (see Fig. 31.2), for journals operating in plain bearings, for cam

Figure 31.1　Hydraulic cylinders

faces, and for rollers of rolling bearings when loads are normal.

Figure 31.2   O-ring packings

0.4μm The surface roughness is used for rapidly rotating shaft bearings, for heavily loaded bearings, for rolls in bearings of ordinary commercial grades, for hydraulic applications, for the bottom of sealing-rings grooves, for journals operating in plain bearings, and for extreme tension members.

0.8μm The surface roughness is normally found on parts subject to stress concentrations and vibrations, for broached holes, gear teeth, and other precision machined parts.

1.6μm This surface roughness is suitable for ordinary bearings, for ordinary machine parts where fairly close dimensional tolerances must be held, and for highly stressed parts that are not subject to severe stress reversals.

3.2μm The surface roughness should not be used on sliding surfaces, but can be used for rough bearing surfaces where loads are light and infrequent, or for moderately stressed machine parts.

6.3μm The appearance of this surface roughness is not objectionable, and can be used on noncritical component surfaces, and for mounting surfaces for brackets, etc.

Because of the highly competitive nature of most manufacturing businesses, the question of finding ways to reduce cost is ever present. A good starting point for cost reduction is in the design of the product. The design engineer should always keep in mind the possible alternatives available to him in making his design. It is often impossible to determine the best alternatives without a careful analysis of the probable production cost. Designing for function, interchangeability, quality, and economy requires a careful study of tolerances, surface roughness, processes, materials, and equipment.

A comprehensive study of the principles of interchangeability is essential for a thorough understanding and full appreciation of low-cost production techniques. Interchangeability is the key to successful production regardless of quantity. Details of all parts should be surveyed carefully to assure not only inexpensive processing but also rapid, easy assembly and maintenance.

Words and Expressions

size specification 规格,规格尺寸

deviation [ˌdiːviˈeiʃən] *n*. 偏差,差别

unilateral [ˈjuːniˈlætərəl] *a*. 单方面,单边的,片面的

bilateral [baiˈlætərəl] *a*. 有两面的,双边的

clearance fit 间隙配合

interference fit 过盈配合

transition fit 过渡配合

interior [inˈtiəriə] *a*. 内部的,里面的,内心的,本质的; *n*. 内部,里面

strut [strʌt] *n*. 支柱,支杆,竖直构件支柱

packing [ˈpækiŋ] *n*. 包装,组装,填充物,密封垫,密封件

coarse [kɔːs] *a*. 粗的,粗糙的,未加工的,不精确的

infrequent [inˈfriːkwənt] *a*. 不常见的,很少发生的,不寻常的

bracket [ˈbrækit] *n*. 托架,轴承架,支座

piston [ˈpistən] *n*. 活塞,柱塞

shaft bearing 轴承

stress concentration 应力集中

objectionable [əbˈdʒekʃnəbl] *a*. 该反对的,不能采用的,不适宜的,不好的

mounting [ˈmauntiŋ] *n*. 安装,安置,装配,固定,配件

alternative [ɔːlˈtəːnətiv] *a*. 交替的,替换的; *n*. 比较方案,替换物

survey [ˈsəːvei] *v*. 观察,测量,综述,全面的观察

32 Fundamentals of Manufacturing Accuracy

Manufacturing can be defined as the transformation of raw materials into useful products through the use of the easiest and least-expensive methods. It is not enough, therefore, to process some raw materials and obtain the desired product. It is, in fact, of major importance to achieve that goal through employing the easiest, fastest, and most efficient methods. If less efficient techniques are used, the production cost of the manufactured part will be high, and the part will not be as competitive as similar parts produced by other manufacturers. Also, the production time should be as short as possible to enable capturing a larger market share.

Modern industries can be classified in different ways. There include classification by process, classification by product, and classification based on the production volume and the diversity of products. The classification by process is exemplified by casting industries, stamping industries, and the like. When classifying by product, industries may belong to the automotive, aerospace, and electronics groups. The third method, i. e., classification based on production volume, identifies three main distinct types of production, mass, job shop, and moderate.

Mass production is characterized by the high production volume of the same (or very similar) parts for a prolonged period of time. The typical example of mass-produced goods is automobiles.

Job-shop production is based on sales orders for a variety of small lots. Each lot may consist of 20 up to 200 or more similar parts, depending upon the customers' needs. It is obvious that this type of production is most suitable for subcontractors who produce varying components to supply various industries.

Moderate production is an intermediate phase between the job-shop and the mass-production types. This type of production is gaining popularity in industry because of an increasing market demand for customized products.

A very important fact of the manufacturing science is that it is almost impossible to obtain the desired basic size when processing a workpiece. This is actually caused by the inevitable, though very slight, inaccuracies inherent in the machine tool as well as by various complicated factors like the elastic deformation and recovery of the workpiece and/or the fixture, temperature effects during processing, and sometimes the skill of the operator. Since it is very difficult to analyze and completely eliminate the effects of these factors, it is more feasible to establish a permissible deviation from the basic size that would not affect the proper functioning of the manufactured part in a detrimental way. The deviations from the basic size to each side (i.e., positive or negative) determine the high and the low limits, respectively, and the difference between those two limits of size is called the tolerance. The tolerance is an absolute value without a sign.

The tolerance may be indicated by a basic size and tolerance symbol. A tolerance symbol consists of the combination of the IT grade number and the tolerance position letter. A particular symbol specifies the upper- and lower-limit dimensions for a part. The tolerance sizes are thus defined by the basic size of the part followed by the symbol composed of a letter and a number (see Fig. 32.1). It is obvious that smaller tolerances require the use of high-precision machine tools in manufacturing the parts and therefore increase production costs.

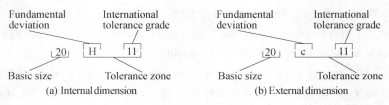

Figure 32.1   Metric tolerance symbol

Before two components are assembled together, the relationship between the dimensions of the mating surfaces must be specified. This actually determines the degree of tightness or freedom for relative motion between the mating surfaces. There are basically three types of fits, namely, clearance fit, transition fit, and interference fit. In all cases of clearance fit, the upper limit of the shaft is always smaller than the lower limit of the mating hole. This is not the case in interference fit, where the lower limit of the shaft is always larger than the upper limit of the hole. The transition fit, as the name suggests, is an intermediate fit. According to ISO, the internal enveloped part is always referred to as the shaft, whereas the surrounding surface is referred to as the hole. Accordingly, from the fits point of view, a key is referred to as the shaft and the keyway as the hole (see Fig. 32.2).

Figure 32.2 Keys and keyways

There are two ways for specifying and expressing the various types of fits, the shaft-basis and the hole-basis systems. The location of the tolerance zone with respect to the zero line is indicated by a letter, which is always capital for holes and lowercase for shafts, whereas the tolerance grade is indicated by a number, as previously explained. Therefore, a fit symbol can be $\phi 30H7/h6$, $\phi 50F6/g5$, or any other similar form.

When the service life of an electric bulb is over, all you do is buy a new one and replace the bulb. This easy operation would not be possible without two main concepts, interchangeability and standardization. Interchangeability means that identical parts must be interchangeable, i.e., able to replace each other, whether during assembly or subsequent maintenance work. As you can easily see, interchangeability is achieved by establishing a permissible tolerance, beyond which any further deviation from the basic size of the part is not allowed.

Words and Expressions

capture ['kæptʃə] *n.*; *v.* 俘获,夺取,赢得,引起

stamping industry 冲压作业
automotive group 汽车集团
types of production 生产类型
mass production 大量生产，大批生产
job-shop production 单件、小批生产
moderate production 中批生产
prolonged *a.* 持续很久的，长期的，长时间的
subcontractor [ˈsʌbkənˈtræktə] *n.* 第二次转包的工厂，转包人，小承包商
customized product 定制产品
(be) inherent in 为……所固有，是……的固有性质
deviation [ˌdiːviˈeiʃən] *n.* 偏离，偏移，差异，误差，偏差
basic size 基本尺寸，公称尺寸，规定尺寸
fundamental deviation 基本偏差
international tolerance grade 国际公差(IT)等级，标准公差等级
fit symbol 配合符号
tightness [ˈtaitnis] *n.* 紧密性，松紧度，密封性
envelop [inˈveləp] *v.* 包装，包围，包络
internal enveloped part 被包容件
shaft-basis system (公制)基轴制，采用英制单位时为 basic shaft system
hole-basis system (公制)基孔制，采用英制单位时为 basic hole system
capital letter 大写字母
lowercase letter 小写字母
tolerance zone 公差带
zero line 零线(公差与配合中确定偏差的一条基准直线)
standardization [ˌstændədaiˈzeiʃən] *n.* 标准化，规格化，标定，校准

33　Product Drawings

    Manufacturing companies are established to produce one or more products. These products are completely defined through documents referred to as product drawings. The dimensions and specifications found on the product drawings will ensure part interchangeability and a reliable level of designed performance. Product drawings normally include detail drawings and assembly drawings.

    A detail drawing is a dimensioned, multiview drawing of a single part, describing the part's shape, size, material, and surface roughness, in sufficient detail for the part to be manufactured based on the drawing alone. Most parts require three views for complete shape description.

    Assembly drawings are used to describe how parts are put together, as well as the function of the entire unit; therefore, complete shape description is not important. The views chosen should describe the relationships of parts, and the number of views chosen should be the minimum necessary to describe the assembly.

    All parts must interact with other parts to some degree to yield the desired function from a design. Before detail drawings of individual parts are made the designer must thoroughly analyze the assembly drawing to ensure that the parts fit property with mating parts, that the correct tolerances are applied, that the contact surfaces are properly machined, and that the proper motion is possible between the parts.

    The inch is the basic unit of the English system, and virtually all shop drawings in the U.S. are dimensioned in inches.

    The millimeter is the basic unit of the metric system. Metric abbreviation (mm) after the numerals is omitted from dimensions because the SI symbol near the title block indicates that all units are metric.

    In the U.S., some drawings carry both inch and millimeter dimensions, usually the dimensions in parentheses or brackets are millimeters. The units may also appear as millimeters first and then be converted and shown in brackets as inches. Converting from one unit to the other results in fractional round-off errors. And explanation of the primary unit system for each drawing should be noted in the title block.

Title Blocks In practice, title blocks usually contain the title of the drawing or part name, drafter, date, scale, name of the company, and drawing number. Other information, such as checkers, and materials, also may be given. Any changes or modifications added after the first version to improve the design is shown in the revision blocks. The title block is generally located in the lower right-hand corner of the drawing sheet.

The scale is the ratio between the linear dimensions of the drawing of an object and the actual dimensions. Whatever scale is used, the dimensions on the drawing indicate the true size of the object, not of the view. Various scales may be used for different drawings in a set of product drawings.

Depending on the complexity of the project, a set of product drawings may contain from one to more than a hundred sheets. Therefore, giving the number of each sheet and the total number of sheets in the set on each sheet is important (for example, sheet 2 of 6, sheet 3 of 6, and so on).

Part Names and Numbers Give each part a name and number, usually using letters and numbers 1/8-in. (3mm) high. Place part numbers near the views to which they apply, so their association will be clear.

Parts List The part numbers and part names in the parts list correspond to those given to each part depicted on the product drawings. In addition, the number of identical parts required is given along with the material used to make each part.

Up to 1900, drawings everywhere were generally made in what is called first-angle projection. In the first-angle projection, the top view is placed under the front view, the left-side view is placed at the right of the front view, and so on. Today, third-angle projection is in common use in the USA, Canada, and the UK, but first-angle projection is still used throughout much of the world. In the third-angle projection, the top view is placed above the front view, the right-side view is placed at the right of the front view, the left-side view is placed at the left of the front view.

Actually, the only difference between first-angle and third-angle projection is the arrangement of views. Confusion and manufacturing errors may result when the user reading a first-angle drawing thinks it is a third-angle drawing, or vice versa. To avoid misunderstanding, projection symbols, shown in Fig. 33.1, have been developed to distinguish between first-angle and third-angle projections on drawings. On drawings where the possibility of confusion is anticipated, these symbols may appear in or near the title block.

(a) First-angle projection   (b) Third-angle projection

Figure 33.1   Projection symbols

Product drawings are legal contracts that document the design details and specifications as directed by the design engineer. Therefore drawings must be as clear, precise, and thorough as possible. Revisions and modifications of a project at the time of production are much more expensive than when done in the preliminary design stages. To be economically competitive drawings must be as error-free as possible.

People who check drawings must have special qualifications that enable them to identify errors and to suggest revisions and modifications that result in a better product at a lower cost. Checkers inspect assembly or detail drawings for correctness and soundness of design. In addition, they are responsible for the drawing's completeness, quality, and clarity.

In addition to the individual revision records, drafters should keep a log of all changes made during a project. As the project progresses, the drafter should record the changes, dates, and people involved. Such a log allows anyone reviewing the project in the future to understand easily and clearly the process used to arrive at the final design.

Words and Expressions

drawing 图样(根据投影原理、标准或有关规定,表示工程对象,并有必要的技术说明的图),图
product drawing 产品图样,产品图纸
interchangeability [ˌintə(ː)ˌtʃeindʒə'biliti] *n*. 互换性
assembly drawing 装配图
detail drawing 零件图,详图,细部图
dimension [di'menʃən] *v*. 标注尺寸; *n*. 尺寸
multiview drawing 含有多向视图的图样,含有多个视图的图样
view [vjuː] *n*. 视图,外形图
specification [ˌspesifi'keiʃən] *n*. 规格,说明书,规范,规格,技术要求
consolidate [kən'sɔlideit] *v*. 巩固,统一
fit 配合,适合
mating part 配合件
tolerance ['tɔlərəns] *n*. 公差; *v*. 给(机器部件等)规定公差
machine [mə'ʃiːn] *n*. 机器; *v*. 机械加工
thorough ['θʌrə] *a*. 完整的,彻底的,全面的
English system (以英尺、磅、秒、加仑等为基本单位的)英制
metric system 公制,米制(以米、千克、秒等为基本单位)
shop drawing 车间加工图,制造图样
abbreviation [əˌbriːvi'eiʃən] *n*. 缩写,缩写词
parenthesis [pə'renθisis] *n*. (*pl*. parentheses)圆括号
bracket ['brækit] *n*. 方括弧,中括号
fractional ['frækʃənl] *a*. 分数的,小数的
round-off error 舍入误差,四舍五入产生的误差
explanation [ˌeksplə'neiʃən] *n*. 说明,解释
title block 标题栏(由名称及代号区、签字区、更改区和其它区组成的栏目)
scale [skeil] *n*. 比例
drawing number 图号,图样代号
drawing sheet 图纸
checker ['tʃekə] *n*. 审核人员

parts list 零件明细表,明细栏(包含零件序号、代号、名称、规格、数量、材料等内容的表格),也可写为 item block 或 item list

soundness ['saundnis] *n*. 完善,无缺陷

part number 零件序号

review [ri'vju:] *v*.;*n*. 检查,评审,评阅,审核

first-angle projection 第一角投影,第一角画法

top view 俯视图(由上向下投影得到的视图。在第三角投影中,俯视图配置在主视图的上方)

front view 主视图(由前向后投影得到的视图)

left-side view 左视图,也可写为 left view

right-side view 右视图,也可写为 right view

projection symbol 投影识别符号,投影符号

log [lɔg] *n*. 日志,(工程、试验等的)工作记录

keep a log of 将……记录下来

34 Sectional Views

Sectional views, frequently called sections, are used to show internal construction of an object that is too complicated to be shown clearly by regular views containing many hidden lines. To produce a sectional view, a cutting plane is assumed to be passed through the part, and then removed.

Section lining. Section lining can serve a double purpose. It indicates the surface that has been cut and makes it stand out clearly, thus helping the observer to understand the shape of the object. Section lining may also indicate the material from which the object is made, when the lining symbols shown in Fig. 34.1 are used.

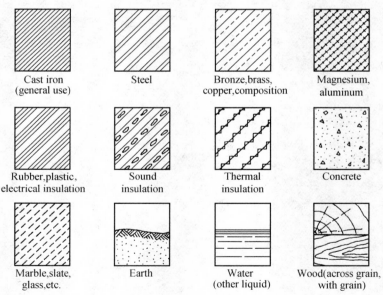

Figure 34.1 ANSI standard section lining symbols for various materials

Full Sections. When the cutting plane extends entirely through the object in a straight line and the front half of the object is imagined to be removed, a full section is obtained. This type of section is used for both detail and assembly drawings. When the section is on an axis of symmetry, it is not necessary to indicate its location. Figure 34.2 shows a full section with cutting plane omitted.

Half Sections. If a cutting plane passes half way through an object, the result is a half section. The half section is used when you wish to show both internal construction and the exterior view of an object in the same view. This type of section is most effective when the object is symmetric or nearly so. Therefore, one half of the object illustrates internal construction, and the other half shows an external view with hidden lines omitted. The half section is visualized by imagining that one quarter of the object has been removed. Figure 34.3 shows the position of the cutting plane in the top view and the resulting half section in the front view. The section view and the external view are separated by a centerline. The hidden lines in the external view portion have been omitted according to conventional practice, although they could be added if needed to clarify the drawing.

The half section is not widely used in detail drawings because of difficulties in dimensioning internal shapes that are shown in part only in the sectioned half (Fig. 34.3). The greatest usefulness of the half section is in assembly drawing, in which it is often necessary to show both internal and external constructions on the same view.

Broken-Out Sections. Where a sectional view of only a portion of the object is needed, broken-out sections may be used (Fig. 34.4). An irregular break line is used to show the extent of the section.

Offset Sections. In order to include features that are not in a straight line, the cutting plane may be offset, so as to include several planes. Such a section is called an offset section. Figure 34.5 shows a typical use of an offset section.

Aligned Sections. To include in a section certain angled features, the cutting plane may be bent to pass through those features. The features cut by the cutting plane are then imagined to be revolved into a plane. Figure 34.6 is an example of aligned section.

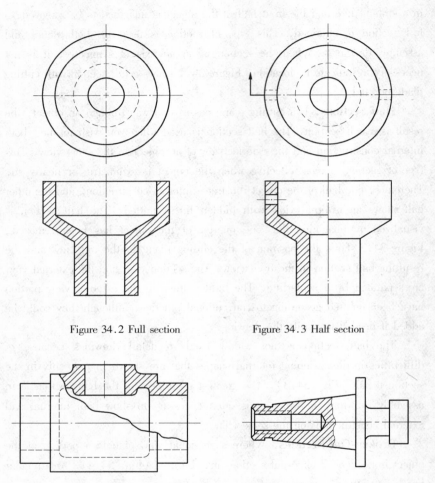

Figure 34.2 Full section Figure 34.3 Half section

Figure 34.4 Broken-out sections

Placement of Sectional Views. Whenever practical, sectional views should be projected perpendicularly to the cutting plane and be placed in the normal position for third-angle projection. When the preferred placement is not practical, the sectional view may be removed to some other convenient position on the drawing, but it must be clearly identified, usually by two identical capital letters, one at each end of the line. Normally, the sections are labeled alphabetically, beginning with A-A, then B-B, and so on (Fig. 34.7). Note that view C-C is not a sectional view since the cutting plane is external to the object.

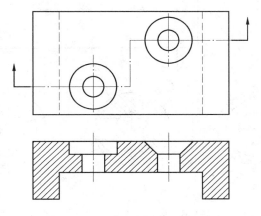

Figure 34.5   Offset section

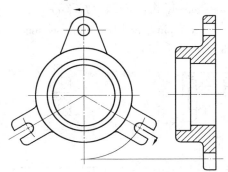

Figure 34.6   Aligned section

Section Lining on Assembly Drawings. The cast iron symbol (evenly spaced section lines) is recommended for most assembly drawings. The section lining should be drawn at an angle of 45° with the main outline of the view. On adjacent parts, the section lines should be drawn in the opposite direction, as shown in Fig. 34.8.

Shafts, bolts, nuts, screws, rivets, pins, washers, and gear teeth, should not be sectioned except that a broken-out section of the shaft may be

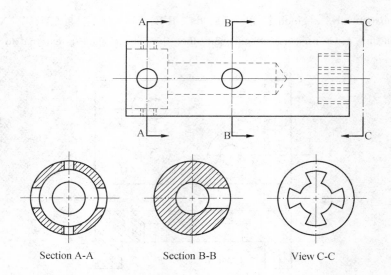

Figure 34.7 Detail drawing having two sectional views

used to describe more clearly the key, keyseat, or pin (Fig. 34.8). If the shafts, bolts, nuts, rivets, pins, and keys were sectioned, the drawing would be confusing and difficult to read.

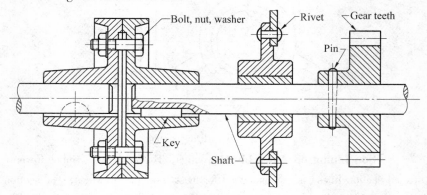

Figure 34.8 Section lining on an assembly drawing

Words and Expressions

view 视图,外形图

sectional view 剖视图,剖面图
section *n*.; *v*. 剖视,剖面,剖面图,剖视图
internal construction 内部结构
hidden line 虚线
cutting plane 剖切面
section lining 剖面线,也可写为 **section line**
serve a double purpose 起双重作用
stand out clearly 清晰地显示出
section lining symbol 剖面符号,也可写为 **section line symbol**
general use 通常,一般用途
marble ['mɑ:bl] *n*. 大理石
slate [sleit] *n*. 石板,石片
across grain 横纹,与木纹垂直,横剖面
with grain 顺纹,顺纹理,纵剖面
full section 全剖视图
axis of symmetry 对称轴,对称轴线
half section 半剖视图
symmetric [si'metrik] *a*. 对称的,平衡的;*n*. 对称
centerline ['sentəlain] 中心线
external view 外观图,外形视图
dimension [di'menʃən] *v*. 在……上标出尺寸
broken-out section 局部剖视,也可以写为 **partial section**
break line 波浪线
offset ['ɔ:fset] *a*. 偏移的,横向移动的
offset section 转折剖(相当于原 GB 中的阶梯剖),用几个平行的剖切平面获得的剖视图
a typical use of 一种典型的用法
aligned section 摆正剖(相当于原 GB 中的旋转剖),用几个相交的剖切平面获得的剖视图
other than M 除了 M,M 除外,与 M 不同的
label [leibl] *n*. 标签,标志;*v*. 标注
alphabetically [ˌælfə'betikəli] *ad*. 按字母(表),顺序,按 ABC 顺序

placement of sectional view 剖视图的配置
project [ˈprɔdʒekt] *n.*; *v.* 投射，投影
evenly spaced 间隔相等的，等间隔的，也可写为 **equally spaced**
main outline 主要轮廓线
section line *n.*; *v.* 剖面线，画剖面线
confusing [kənˈfjuːziŋ] *a.* 混乱的，混淆的，令人困惑的
keyseat 键槽，也可以写为 **keyway** 或 **key seat**

35 Nontraditional Manufacturing Processes

The human race has distinguished itself from all other forms of life by using tools and intelligence to create items that serve to make life easier and more enjoyable. Through the centuries, both the tools and the energy sources to power these tools have evolved to meet the increasing sophistication and complexity of mankind's ideas.

In their earliest forms, tools primarily consisted of stone instruments. Considering the relative simplicity of the items being made and the materials that were being shaped, stone was adequate. When iron tools were invented, durable metals and more sophisticated articles could be produced. The twentieth century has seen the creation of products made from the most durable and, consequently, the most difficult-to-machine materials in history. In an effort to meet the manufacturing challenges created by these materials, tools have now evolved to include materials such as alloy steel, carbide, diamond, and ceramics.

A similar evolution has taken place with the methods used to power our tools. Initially, tools were powered by muscles; either human or animal. However as the powers of water, wind, steam, and electricity were harnessed, mankind was able to further extend manufacturing capabilities with new machines, greater accuracy, and faster machining rates.

Every time new tools, tool materials, and power sources are utilized, the efficiency and capabilities of manufacturers are greatly enhanced. However as old problems are solved, new problems and challenges arise so that the manufacturers of today are faced with tough questions such as the following: How do you drill a 2-mm diameter hole 670-mm deep without experiencing taper or runout? Is there a way to efficiently deburr passageways inside complex castings and guarantee 100% that no burrs were missed? Is there a welding process that can eliminate the thermal damage now occurring to my product?

Since the 1940s, a revolution in manufacturing has been taking place that once again allows manufacturers to meet the demands imposed by increasingly sophisticated designs and durable, but in many cases nearly unmachinable, materials. This manufacturing revolution is now, as it has been in the past, centered on the use of new tools and new forms of energy. The result has been the introduction of new manufacturing processes used for material removal, forming, and joining, known today as nontraditional manufacturing processes.

The conventional manufacturing processes in use today for material removal primarily rely on electric motors and hard tool materials to perform tasks such as sawing, drilling, and broaching. Conventional forming operations are performed with the energy from electric motors, hydraulics, and gravity. Likewise, material joining is conventionally accomplished with thermal energy sources such as burning gases and electric arcs.

In contrast, nontraditional manufacturing processes harness energy sources considered unconventional by yesterday's standards. Material removal can now be accomplished with electrochemical reactions, electrical sparks (see Fig. 35.1), high-temperature plasmas, and high-velocity jets of liquids and abrasives. Materials that in the past have been extremely difficult to form, are now formed with magnetic fields, explosives, and the shock waves from powerful electric sparks. Material-joining capabilities have been expanded with the use of high-frequency sound waves and beams of electrons.

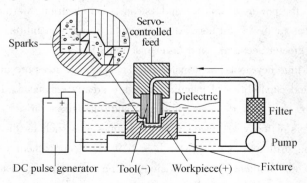

Figure 35.1 Schematic diagram of EDM process

In the past 50 years, over 20 different nontraditional manufacturing processes have been invented and successfully implemented into production. The reason there are such a large number of nontraditional processes is the same reason there are such a large number of conventional processes; each process has its own characteristic attributes and limitations, hence no one process is best for all manufacturing situations.

For example, nontraditional processes are sometimes applied to increase productivity either by reducing the number of overall manufacturing operations required to produce a product or by performing operations faster than the previously used method.

In other cases, nontraditional processes are used to reduce the number of rejects experienced by the old manufacturing method by increasing repeatability, reducing in-process breakage of fragile workpieces, or by minimizing detrimental effects on workpiece properties.

Because of the aforementioned attributes, nontraditional manufacturing processes have experienced steady growth since their introduction. An increasing growth rate for these processes in the future is assured for the following reasons:

(1) Currently, nontraditional processes possess virtually unlimited capabilities when compared with conventional processes, except for volumetric material removal rates. Great advances have been made in the past few years in increasing the removal rates of some of these processes, and there is no reason to believe that this trend will not continue into the future.

(2) Approximately one-half of the nontraditional manufacturing processes are available with computer control of the process parameters. The use of computers lends simplicity to processes that people may be unfamiliar with, and thereby accelerates acceptance. Additionally, computer control assures reliability and repeatability, which also accelerates acceptance and implementation.

(3) Most nontraditional processes are capable of being adaptively-controlled through the use of vision systems, laser gages, and other in-process inspection techniques. If, for example, the in-process inspection system

determines that the size of holes being produced in a product are becoming smaller, the size can be modified without changing hard tools, such as drills.

(4) The implementation of nontraditional manufacturing processes will continue to increase as manufacturing engineers, product designers, and metallurgical engineers become increasingly aware of the unique capabilities and benefits that nontraditional manufacturing processes provide.

Words and Expressions

sophistication [səfisti'keiʃən] *n*. 复杂化,完善,采用先进技术
durable ['djuərəbl] *a*. 耐用的,耐久的,坚固的; *n*. 耐久的物品
runout ['rʌn'aut] *n*. 偏斜,径向跳动
deburr [di'bə:] *v*. 去毛刺,去飞翅
passageway ['pæsidʒ'wei] *n*. 通道,通路
casting ['kɑ:stiŋ] *n*. 铸造,铸件
burr [bə:] *n*. 毛刺
sawing ['sɔ:iŋ] *n*. 锯,锯切,锯开
unconventional ['ʌnkən'venʃənl] *a*. 非传统的,非常规的,不一般的
electrochemical [i'lektrəu'kemikl] *a*. 电化学的
dielectric [ˌdaii'lektrik] *n*. 电介质,工作液(特种加工中指 dielectric fluid)
pulse generator 脉冲发生器,(特种加工中通常称)脉冲电源
EDM 电火花加工(**electrical discharge machining**)
plasmas ['plæzmə] *n*. 等离子
reject [ri'dʒekt] *v*. 排斥,抵制; *n*. 等外品,不合格品,废品
in-process *a*. (加工,处理)过程中的
breakage ['breikidʒ] *n*. 破损,断裂,损坏
fragile ['frædʒail] *a*. 易碎的,易毁坏的
minimize ['minimaiz] *v*. 使……成最小,最小化
aforementioned [ə'fɔ:'menʃənd] *a*. 上述的,前面提到的
thereby ['ðɛə'bai] *ad*. 因此,所以,在那方面,大约
adaptive [ə'dæptiv] *a*. 适合的,适应的,自适应的
adaptive control 自适应控制
implementation [ˌimplimen'teiʃən] *n*. 履行,实现,执行过程

36　Machining of Engineering Ceramics

Engineering ceramic materials have attractive properties: high hardness, high thermal resistance, chemical inertness, and low thermal or electrical conductivity, to name a few. However, these properties make ceramics extremely difficult to machine, whether by abrasive or nonabrasive processes. For the latter, material removal rates must be increased in current processes to improve productivity rates for economic justification. Other requirements include: improved surface roughness, control of geometric features, and reduced equipment costs.

Grinding involves a complex interaction between a number of variables: workpiece material properties, wheel specifications, machine tools selection, and wheel preparation. This interaction, called "grindability," can be quantified in terms of material-removal rate, power or grinding-force required, surface roughness, tolerances, and surface integrity.

During grinding flaws are introduced on the surface that vary from piece to piece, affecting the surface integrity. This translates into a variation in fracture strength. Residual stresses are also generated on the surface and subsurface, due to the mechanical process of chip removal. Defective regions within the surface layer can have linear dimensions between 10 and 100 μm. Such damage often must be removed by finish machining with a finer-grit wheel.

For conventional grinding of ceramics, the selection of the grinding machine and the grinding wheel also influence grindability. In general, only diamond grinding wheels can be used for ceramics. Resin-bonded diamond wheels provide better results than either metal or vitrified-bonded wheels for fired ceramics. Resin-bonded wheels produce lower grinding forces and wear away faster, exposing new cutting edges which help to minimize surface flaw size of the ceramic workpiece.

Since the ceramic's properties play a major role in the final grinding results, their interaction with machining variables must be understood completely to determine costs. Coolant application becomes critical for low thermal-conductivity materials to prevent thermally induced cracks. Porosity, grain size, and microstructure affect surface roughness and surface quality; high porosity contributes to poor surface roughness. Generally, ceramics require higher grinding forces and power, which lead to low wheel life.

Several abrasive techniques do not require expensive diamond wheels and the problems that go with them. Water-jet machining uses a high-velocity fluid jet, either alone or with abrasive particles, to erode material (see Fig. 36.1). The jet can be used in the pulsed, or continuous modes—the latter for most ceramic applications. This type of machining is best for cutting slots and grooves or trepanning large holes.

Figure 36.1 Water jet machining process

There are no heat effects nor large mechanical forces in water-jet machining, so that even the most thermal-shock sensitive, weakest ceramics can be cut without damage. Chatter, vibration, surface distortion, and subsurface damage of the workpiece also are eliminated.

Ultrasonic machining (see Fig. 36.2) is another abrasive process that has certain advantages over conventional grinding. Sometimes called impact grinding, it is a mechanical process that uses a high frequency transducer. As the electrical energy is converted into mechanical motion, a low-amplitude vibration is produced. The vibration is transmitted to the toolholder, which results in a linear mechanical movement of typically 0.02 mm at a rate of 20 000 cycles/sec.

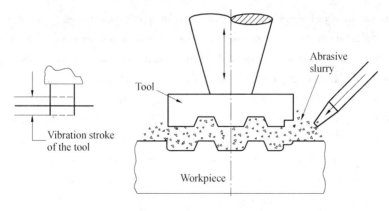

Figure 36.2 Schematic diagram of an ultrasonic machining operation

This action produces a microscopic chipping at a steady penetration rate. A constant flow of abrasive slurry is used under controlled pressure. A 20% to 50% concentration of boron carbide or silicon carbide is the most commonly used abrasives. The slurry is recirculated to provide fresh abrasive and to remove abraded particles. The particle size of the grit determines the surface roughness, the size of the cavity in relation to the tool, and the cutting rate. Vacuum can be used to improve abrasive flow. The tooling is patterned after the configuration to be machined. The toolholder can be amplified to increase the stroke, thus producing faster cutting rates.

Ultrasonic machining can be used to machine almost any hard, brittle material, though it is more effective in materials of 40 HRC or more. These include silicon, glass, quartz, fiber-optic materials, structural ceramics such as SiC, and electronic substrates such as alumina.

Abrasionless machining, such as laser-beam and electron-beam machining (LBM, EBM), is not limited by the hardness of ceramics and therefore offers an alternative to conventional grinding. In LBM, a laser focuses an intense beam of light onto the workpiece to vaporize material. Because the laser beam is mechanically positioned, it is not as fast as EBM. However, unlike EBM, no vacuum chamber is required so workpiece loading is faster, and there is no limit on size. Another advantage is less-expensive equipment. Usually a pulsed mode is used for drilling, while a continuous mode is used for cutting. LBM can machine just about any hard material, including diamond.

Machining rate is controlled by the rate at which material, which is melted and vaporized, is removed by the beam. Removal occurs by thermal convection and beam pressure. The rate may be greatly increased if a gas jet is used to blow away the material. Thermal gradients can cause cracking so that thick ceramics may have to be machined in the unfired state. Conversely, laser-machined ceramics may be stronger than diamond-machined ceramics since heating can produce beneficial residual stresses. Laser power requirements are determined by the amount of stock to be removed; one 150-W laser is sufficient for thin ceramics, while a 15 000-W laser usually is required for thick ceramics.

Words and Expressions

inertness [i'nə:tnis] *n*. 惰性
removal rate 去除速度,体积加工速度
justification [ˌdʒʌstifi'keiʃən] *n*. 证明为正当,辩护,正当的理由
grindability 磨削性,可磨性
chatter ['tʃætə] *n*.; *v*. 颤振,自激振动
vitrify ['vitrifai] *v*. (使)玻璃化,使成玻璃状物质
vitrified bond 陶瓷结合剂
porosity [pə'rɔsiti] *n*. 多孔性,孔隙率,密集气孔
jet [dʒet] *n*. 射流,水流; *v*. 喷出,喷射,射流
erode [i'rəud] *v*. 侵蚀,蚀除
ultrasonic machining 超声加工,超声波加工
transducer [trænz'djuːsə] *n*. 换能器,变换器,传感器
boron ['bɔːrɔn] **carbide** 碳化硼
slurry ['slə:ri] *n*. 悬浮液,膏剂,软膏,磨料粉浆
concentration [ˌkɔnsen'treiʃən] *n*. 集中,浓缩,浓度,密度,集度
pattern ['pætən] *n*. 模型,图形; *v*. 模仿,仿造(after)
convection [kən'vekʃən] *n*. 对流,迁移,传递
gradient ['greidiənt] *n*. 梯度,变化率

37 Mechanical Vibrations

Mechanical vibrations are the oscillatory motions, either continuous or transient, of objects and structures. In some instances they are purposeful and integral to the design of a machine as in a pneumatic drill or a reciprocating engine. In most instances, however, they are incidental or accidental and may impair the normal functioning of a structure or instrument.

Such vibrations enter into all aspects of the mechanical engineering and are therefore of interest to some extent in all fields of engineering science and physics. A knowledge of the fundamentals of mechanical vibrations is indispensable to practitioners of these varied technologies.

Effects of Vibrations in Mechanical Systems There are a number of weighty reasons for the widespread interest in the fundamentals and practical aspects of mechanical vibrations.

One such reason is the possibility of undesirable effects by vibrations on mechanical systems. Any general mechanical system, for example, whole buildings, instruments on the bench in the laboratory, complex mechanical tools on the floor of a workshop, transport vehicles or a human being may be represented by some pattern or form of interconnected mass/spring/damper elements. Since most driving forces $f(t)$ may be deemed to have harmonic components, the possibility of exciting resonances within the over-all system is great. If a resonance is not damped, the displacement of the mass and hence the stretching of the spring element will tend towards infinity. The spring component will fracture and for this reason undamped resonances must be avoided for the protection of equipment and instruments. This applies also when the human body is part of the over-all system which might experience the damaging resonance.

Long-term exposure of a mechanical system to vibrations of frequencies away from resonance can also cause damage through the mechanism of fatigue.

Thus, if a mechanical component such as a spring is subjected to repetitive or cyclical applications of stress levels much lower than the ultimate strength, it will fracture after a large number of repetitions of this stress. Indeed, if the number of cycles of stress is increased, the amplitude of the stress needed eventually to cause fracture becomes lower. The underlying mechanism in fatigue appears to be the gradual unzipping of intermolecular bonds starting from a defect or weakness in the molecular structure.

Another undesirable effect of vibrations is the fact that they can impair the normal functioning of instruments. Thus, if there are vibrations within an electron microscope which magnifies by over $\times\ 10^4$, a blurred image can result. Vibrations in a microtome can result in cuts of different thicknesses. Likewise, many devices in precision engineering and optics cannot tolerate excessive vibrations. Electrical connections can be undone by vibrations.

Unwanted vibrations in a system, furthermore, indicate inefficiency. Energy is wasted in exciting the vibrations instead of being effectively directed to the work of the system.

Another undesirable side-effect of vibrating structures is the generation of audible noise. Such noise can be psychologically annoying to human beings working in the environment and can render normal voice communication impossible. If extreme, noise can irreparably damage human hearing. The most thorough way of suppressing such noise is to reduce or eliminate the vibrations causing it.

Considerable effort is devoted to the measurement and examination of seismic vibrations associated with earthquakes. These measurements are a vital link in providing advance warning and protection to populations against volcanic eruptions with which are associated earth tremors.

Another area of interest in vibration quantification is the so-called planned or preventive maintenance of equipment, particularly rotating machinery. As this type of machinery ages and undergoes wear, the associated unwanted vibrations tend to become greater. Regular vibration measurement can provide in-service indices of the degeneration of the machinery. Repair or replacement can then be carried out before catastrophic failure and at a time convenient for the factory or plant.

The first step in any of these areas of vibration science is to measure the vibrations in question.

Measurement Equipment The most generally used methods of measuring vibrations are electrical. The basic components in such methods are shown schematically in Figure 37.1. The key component is the vibration transducer which produces an electrical voltage or current proportional to some quantity in the mechanical vibration, the displacement, velocity or acceleration. Thereafter, a variety of electronic components can carry out any of a range of standard electronic signal processing steps on the vibration voltage. Typical steps include amplification, attenuation, filtering, differentiation and integration. Then the processed signal is measured with a meter, displayed on an oscilloscope (see Fig. 37.2), recorded on a chart recorder or tape recorder or further processed and analyzed by digital computer.

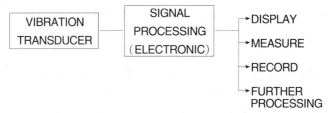

Figure 37.1 Schematic diagram of a system for detecting and measuring vibrations

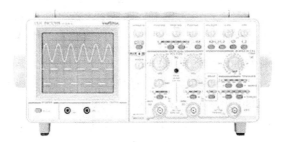

Figure 37.2 An oscilloscope

Words and Expressions

oscillatory ['ɔsileitəri] *a*. 振动的, 振荡的, 摆动的, 摇动的
transient ['trænziənt] *a*. 瞬时的, 瞬态的, 过渡的, 短暂的, 不稳定的
purposeful ['pə:pəsful] *a*. 有目的, 故意的, 果断的
incidental [ˌinsi'dentl] *a*. 偶然发生的, 易发生的, 伴随的
impair [im'pɛə] *v*.; *n*. 削弱, 损害, 减少, 断裂
practitioner [præk'tʃənə] *n*. 专业人员, 开业者
weighty ['weiti] *a*. 重的, 重要的, 有分量的, 有影响的
deem [di:m] *v*. 想, 认为, 相信
harmonic ['hɑ:mənik] *n*. 谐波, 谐波分量, 谐振荡, 谐函数
resonance ['rezənəns] *n*. 共振, 谐振
infinity [in'finiti] *n*. 无限, 无穷大, 无数, 大量
unzip ['ʌn'zip] *v*. 拉开(拉链)
intermolecular [ˌintəmou'lekjulə] *a*. 分子的, 分子组成的
blur [blə:] *v*. 弄脏, 变模糊, 影像位移; *n*. 污迹, 模糊
microtome ['maikrətəum] *n*. 切片刀, (薄片)切片机
undo ['ʌn'du:] (undid, undone) *v*. 拆开, 松开, 使恢复原状, 取消
side-effect *n*. 副作用, 边界效应
irreparably [i'repərəbli] *ad*. 不能修理地, 不能弥补地, 无可挽救地
suppress [sə'pres] *n*. 压制, 扑灭, 抑制, 制止, 排除, 隐蔽
volcanic [vɔl'kænik] *a*. 火山的, 由火山作用所引起的
eruption [i'rʌpʃən] *n*. 喷出, 爆发, 萌出, 喷出物
tremor ['tremə] *n*. 振动, 颤抖, 地震
indices ['indisi:z] index 的复数, 索引, 指数, 系数, 指标, 符号
catastrophic [ˌkætə'strɔfik] *a*. 大变动的, 灾难性的, 突发性的
amplification [ˌæmplifi'keiʃən] *n*. 放大(系数, 作用), 增强, 扩大
attenuation [əˌtenju'eiʃən] *n*. 衰减(现象, 量), 减弱, 降低
filter ['filtə] *n*. 过滤器, 滤波; *v*. 过滤, 滤波
oscilloscope ['ɔsiləskəup] *n*. 示波器, 示波仪
filtering *n*. 过滤, 滤除, 滤波; *a*. 过滤的

38 Definitions and Terminology of Vibration

All matter—solid, liquid and gaseous—is capable of vibration, e.g. vibration of gases occurs in tail ducts of jet engines causing troublesome noise and sometimes fatigue cracks in the metal. Vibration in liquids is almost always longitudinal and can cause large forces because of the low compressibility of liquids, e.g. pipes conveying water can be subjected to high inertia forces (or "water hammer") when a valve or tap is suddenly closed. Excitation forces caused, say, by changes in flow of fluids or out-of-balance rotating or reciprocating parts, can often be reduced by attention to design and manufacturing details. A typical machine has many moving parts, each of which is a potential source of vibration or shock-excitation. Designers face the problem of compromising between an acceptable amount of vibration and noise, and costs involved in reducing excitation.

The mechanical vibrations dealt with are either excited by steady harmonic forces (i.e. obeying sine and cosine laws in cases of *forced* vibrations) or, after an initial disturbance, by no external force apart from gravitational force called weight (i.e. in cases of *natural* or free vibrations). Harmonic vibrations are said to be "simple" if there is only one frequency as represented diagrammatically by a sine or cosine wave of displacement against time.

Vibration of a body or material is periodic change in position or displacement from a *static-equilibrium position*. Associated with vibration are the interrelated physical quantities of acceleration, velocity and displacement—e.g. an unbalanced force causes acceleration ($a = F/m$) in a system which, by resisting, induces vibration as a response. The vibratory or oscillatory motion may be classified broadly as (a) transient; (b) continuing or steady-state; and (c) random.

Transient Vibrations die away and are usually associated with irregular disturbances, e.g. shock or impact forces, rolling loads over bridges, cars driven over pot holes—i.e. forces which do not repeat at regular intervals. Although transients are temporary components of vibrational motion, they can cause large amplitudes initially and consequent high stress but, in many cases, they are of short duration and can be ignored leaving only steady-state vibrations to be considered.

Steady-State Vibrations are often associated with the continuous operation of machinery and, although periodic, are not necessarily harmonic or sinusoidal. Since vibrations require energy to produce them, they reduce the efficiency of machines and mechanisms because of dissipation of energy, e.g. by friction and consequent heat-transfer to surroundings, sound waves and noise, stress waves through frames and foundations, etc. Thus, steady-state vibrations always require a continuous energy-input to maintain them.

Random Vibration is the term used for vibration which is not periodic, i.e. has no cyclic basis and is not regularly repetitive.

In the following paragraphs certain terms and definitions used in vibration analysis are made clear—several of which are probably known to engineering students already.

Period, Cycle, Frequency and Amplitude A steady-state mechanical vibration is the motion of a system repeated after an interval of time known as the *period*. The motion completed in any one period of time is called *a cycle*. The number of cycles per unit of time is called the *frequency*. The maximum displacement of any part of the system from its static-equilibrium position is the *amplitude* of the vibration of that part—the total travel being twice the amplitude. Thus, "amplitude" is not synonymous with "displacement" but is the maximum value of the displacement from the static-equilibrium position.

Natural and Forced Vibration A natural vibration occurs without any external force except gravity, and normally arises when an elastic system is displaced from a position of stable equilibrium and released, i.e. natural vibration occurs under the action of restoring forces inherent in an elastic system, and natural frequency is a *property* of the system.

A forced vibration takes place under the excitation of an external force (or externally-applied oscillatory disturbance) which is usually a function of time, e.g. in unbalanced rotating parts, imperfections in manufacture of gears and drives. The frequency of forced vibration is that of the exciting or impressed force, i.e. the forcing frequency is an arbitrary quantity independent of the natural frequency of the system.

Resonance Resonance describes the condition of maximum amplitude. It occurs when the frequency of an impressed force coincides with, or is near to a natural frequency of the system. In this critical condition, dangerously large amplitudes and stresses may occur in mechanical systems but, electrically, radio and television receivers are designed to respond to resonant frequencies. The calculation or estimation of natural frequencies is, therefore, of great importance in all types of vibrating and oscillating systems. When resonance occurs in rotating shafts and spindles, the speed of rotation is known as the *critical speed*. Hence, the prediction and correction or avoidance of a resonant condition in mechanisms is of vital importance since, in the absence of damping or other amplitude-limiting devices, resonance is the condition at which a system gives an infinite response to a finite excitation.

Damping Damping is the dissipation of energy from a vibrating system, and thus prevents excessive response. It is observed that a natural vibration diminishes in amplitude with time and, hence, eventually ceases owing to some restraining or damping influence. Thus, if a vibration is to be sustained, the energy dissipated by damping must be replaced from an external source.

The dissipation is related in some way to the relative motion between the components or elements of the system, and is caused by frictional resistance of some sort, e.g. in structures, internal friction in material, and external friction caused by air or fluid resistance called "viscous" damping if the drag force is assumed proportional to the relative velocity between moving parts. One device assumed to give viscous damping is the "dashpot" which is a loosely-fitting piston in a cylinder so that fluid can flow from one side of the piston to the other through the annular clearance space. A dashpot cannot store energy but can only dissipate it.

Words and Expressions

terminology [tə:mi'nɔlədʒi] *n*. 专门名词,术语
duct [dʌkt] *n*. 导管,喷管,输送管
longitudinal [lɔndʒi'tju:dinl] *a*. 长度的,纵向的,轴向的
compressibility [kəmpresi'biliti] *n*. 可压缩性
excitation [ˌeksi'teiʃən] *n*. 刺激,扰动,干扰,激励,激振
out-of-balance 不平衡,失去平衡
shock-excitation *n*. 震激,冲击激励
static-equilibrium position 静平衡位置
interrelate [ˌintəri'leit] *v*. 相互有关,互相联系
die away 衰减,逐渐消失
pot hole 坑洞
disturbance [dis'tə:bəns] *n*. 扰动,扰乱,紊乱,故障
steady-state vibration 稳态振动
sinusoidal [sainə'sɔidəl] *a*. 正弦波的,正弦曲线的
dissipation [ˌdisi'peiʃən] *n*. 消散,浪费,消耗
random vibration 随机振动
period, cycle, frequency and amplitude 周期、循环、频率和振幅
natural vibration 自由振动,亦可写为 free vibration
forced vibration 强迫振动,受迫振动
natural frequency 固有频率
critical speed 临界速度,临界转速
damping ['dæmpiŋ] *n*.; *a*. 阻尼,减振,缓冲,衰减,抑制
diminish [di'miniʃ] *v*. 减少,缩小,递减,削弱
viscous ['viskəs] *a*. 粘性的,粘稠的
dashpot ['dæʃpɔt] *n*. 减振器,缓冲器,阻尼器
annular clearance 环状间隙
loosely fitting 松弛配合

39 Residual Stresses

A residual stress is one that exists without external loading or internal temperature differences on a structure or machine. It is usually a result of manufacturing or assembling operations. When the structure or machine is put into service, the service loads superimpose stresses. If the residual stresses add to the service-load stresses, they are detrimental; if they subtract from the service-load stresses they are beneficial.

Only a few examples of detrimental residual stresses will be given here. One, in the assembly of machinery, occurs when two shafts are not in line, and they are forced into connection by rigid couplings. The resulting stresses in the shafts become reversing stresses when the shafts are rotated. The correction, when perfect alignment cannot be economically attained, as is frequently the case, is to use flexible couplings of a type necessary for the degree of misalignment.

Detrimental residual stresses commonly result from differential heating or cooling. A weld is a common example. The weld metal and the areas immediately adjacent are, after solidification, at a much higher temperature than the main body of metal. The natural contraction of the metal along the length of the weld is partially prevented by the large adjacent body of cold metal. Hence residual tensile stresses are set up along the weld.

A general rule is that the "last to cool is in tension," although there is an exception if certain transformations of microstructure occur. Methods for minimizing or reversing these stresses include annealing for stress relief and hammer or shot peening of the weakened surface. Annealing requires heating mild steel to $600 \sim 650℃$, some alloy steels to $870℃$, then holding for a period of time, followed by slow cooling. Some preheating of the parts to be joined may minimize the tensile stresses in welds.

A thin but highly effective surface layer of compressive stress may be induced by roller burnishing, ballizing, and peening processes. It is seen that these processes work-harden an outer layer, thus causing compressive stresses to remain, together with minor tensile stresses in adjacent interior layers. Since the compressive layer is readily obtained all around, these processes are suitable for reversing loads and rotating components where the stress varies between tension and compression. The processes must be carefully controlled in respect to roller pressures and feeds, shot size and speed, etc., for which extensive information is available in engineering books and periodicals.

In roller burnishing process, the surface of the component is cold worked by a hard and highly polished roller or set of rollers. This process is used on various flat, cylindrical, or conical surfaces (Fig. 39.1). Roller burnishing reduces surface roughness by removing scratches and tool marks and induces beneficial compressive surface residual stresses for improved fatigue life. Thread rolling of bolts and screws has long been part of a forming process that not only forms but strengthens the threads by deformation and grain flow around the roots and by inducing compressive residual stresses.

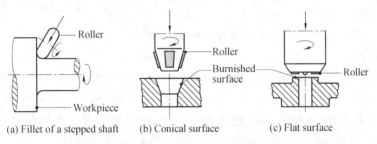

Figure 39.1   Burnishing tools and roller burnishing

Ballizing is a manufacturing process of forcing a hard, tungsten carbide or AISI 52100 steel, slightly oversize ball through a hole in a plate, bushing, or tubing to give the holes final size and a low surface roughness. The length of the hole may be from 1/20 to 10 times its diameter. The machine is often set up for a high production of small parts with unskilled labor. An incidental result is that the process increases hardness, hence wear resistance, and induces around the hole a compressive residual stress that is usually advantageous, as in roller-chain links. The links are highly stressed in pulsating tension with a concentration of the stress at and near the hole surfaces. With the compressive stress from ballizing, the net tensile stress in service is decreased, and failure is minimized.

Peening is the most widely used method for prestressing by mechanically induced yielding. By the impact of rounded striking objects, the surface is deformed in a multitude of shallow dimples, which in trying to expand put the surface under compression.

Shot peening is done on steels by the high-velocity impingement of small, round, steel or chilled cast-iron shot with diameters from 0.2 to 5 mm. The compressed layer has a depth up to 1.25 mm, less than with hammer peening, but roughly proportioned to the shot size used and its velocity. The residual stress produced is about half of the yield strength of the strain hardened region.

Shot peening is extensively used because it may be applied with minimum cost to most metals and shapes, except some interior ones. On soft metals, glass beads may be used. Helical springs are commonly shot peened, with up to a 60% increase in allowable stress under pulsating loads. Part of the improvement may be due to the removal of the weakening longitudinal scratches left from the wire-drawing operation. Similarly, coarse-machined and coarse-ground surfaces are smoothed and improved by shot peening, which may be a more economical method than producing a final finish by machining or grinding. Peening is not used on bearing and other closely fitting surfaces where high precision is required. A final grinding for accuracy after peening would remove part or all of the residual stress. Machines are available for the automatic and continuous peening of small- and medium-size parts moving on a conveyor or turntable through the blast.

Laser shock peening (LSP), also called laser shot peening, is a new surface strengthening technology. It uses laser-induced, high-energy shock wave to induce compressive residual stresses on the impacted area of metal. Laser intensities necessary for this process are on the order of 100 to 300 J/cm^2

and have a pulse duration of about 30 nanoseconds. This surface-treatment process produces compressive residual stress layers that are typically 1 mm deep. Laser shock peening has been applied successfully and reliably to jet-engine fan blades and to materials such as titanium, nickel alloys, and steels for improved fatigue resistance, wear resistance and corrosion resistance.

Words and Expressions

prestress [pri:'stres] *v.*; *n.* 预加应力于,施加预应力
solidification [sə‚lidifi'keiʃən] *n.* 凝固(作用),固化
elevated ['eliveitid] *a.* 高架的,升高的,提高的
microstructure ['maikrəstrʌktʃə] *n.* 微观结构,显微组织
transformation [‚trænsfə'meiʃən] *n.* 变换,变化,转换
peening ['pi:niŋ] *n.* 用锤尖敲击,喷丸硬化,喷射(加工硬化法)
anneal [ə'ni:l] *v.*; *n.* 退火,(加热)缓冷,逐渐冷却
preheat ['pri'hi:t] *v.* 预热,预先加热
roller burnishing 滚压
ballizing ['bɔ:laiziŋ] *n.* 挤孔(用硬质钢球对孔进行挤压加工),球推压法
fillet ['filit] *n.* 圆角,过渡圆角
burnished surface 滚压过的表面
AISI = American Iron and Steel Institute 美国钢铁学会
52100 steel 一种高碳铬轴承钢
decarburization [di:‚kɑ:bjuri'zeiʃən] *n.* 脱碳作用
dimple ['dimpl] *n.* 凹痕,坑,表面微凹,波纹
impingement [im'pindʒmənt] *n.* 碰撞,冲击,打击,侵入,冲突
chilled [tʃild] *a.* 已冷的,冷却的,冷淬的,冷硬的,冷铸的
pulsating ['pʌlseitiŋ] *a.*; *n.* 脉动(的),脉冲的,片断的
fitting surface 配合面
turntable ['tə:nteibl] *n.* 转台,回转台,回转机构
blast [blɑ:st] *n.*; *v.* 爆炸,冲击,喷砂,喷丸
laser shock peening 激光冲击强化
pulse duration 脉冲持续时间,脉冲宽度,脉宽
jet engine 喷气发动机

40 Laser-Assisted Machining and Cryogenic Machining Technique

To meet the demand for production and precision, researchers and equipment builders are looking outside the bounds of conventional milling, drilling, turning, and grinding. New cutting and machining processes continue to emerge along with methods for modifying the conventional techniques.

Laser-assisted machining. Laser-assisted machining is a process that has long been researched and is now ready to come out of the lab. In this operation, a laser beam is projected onto the part through fiber optics or some other optical-beam delivery unit, just ahead of the tool (as shown in Fig. 40.1). During a cut, laser power is varied to match the profile being cut. CO_2 lasers are usually used with power levels from 200 to 500 W. The laser-induced heat softens the workpiece and makes it easier to cut.

 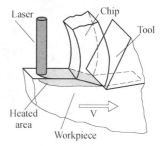

Figure 40.1 Laser-assisted machining process

This process can be used for very difficult-to-machine materials such as superalloy or ceramics. For example, with such process you can cut through ceramics like butter. It also offers a low surface roughness, usually 0.5 μm in Ra or lower.

Because the heating laser beam is tightly focused, heating is localized around the actual cutting zone. Heat is carried off in the chips and there are no changes in the physical properties of the material cut due to heat.

There are two problems with the process. There is a physical problem initially of establishing the cutting parameters (laser power and focus, cutting speeds and feeds, etc.) and the psychological problem of convincing people it works well and has many advantages.

Cryogenic machining technique. An unique use of liquid nitrogen is offering a modification of traditional turning to improve manufacturing operations. Using -195℃ liquid nitrogen as a coolant as a means to lower the heat in the cutting zone offers a number of advantages. The cryogenic coolant system delivers a jet of -195℃ liquid nitrogen directly to the insert during turning operations (as shown in Fig. 40.2). This raises insert hardness, which significantly reduces the thermal softening effect that an insert may experience as a result of hard turning's inherent high cutting temperatures. The steep temperature gradient between the chip/tool interface and insert body also helps remove heat from the cutting zone.

Figure 40.2 Cryogenic machining process

Cubic boron nitride (CBN) and polycrystalline cubic boron nitride (PCBN) inserts have traditionally been the tools of choice for hard-turning applications. But, they are considered too costly for many operations. The cryogenic machining technique also allows greater use of low-cost ceramic inserts (see Fig. 40.3) for hard-turning operations.

CBN and PCBN ceramic inserts tend to wear unevenly and are prone to fracturing when hard turning dry or with water or oil-based coolants. Increased fracture toughness resulting from low-temperature liquid nitrogen cooling provides more predictable, gradual flank wear for ceramic inserts, as well as increasing cutting speeds up to 200%. This predictable flank wear also allows alumina ceramic inserts to be used in critical finishing operations.

Figure 40.3　Ceramic inserts and indexable turning tools

The liquid nitrogen may be stored in a small, dedicated cylinder near a machine, or in a supply tank that would serve multiple machines. Programming is similar to a traditional coolant delivery system. A flexible liquid nitrogen line attaches to a lathe's turret via a rotational coupling. This line feeds a delivery nozzle clamped to the tool, which directs the liquid nitrogen to the insert tip.

Nitrogen is a safe, noncombustible, and noncorrosive gas. It quickly evaporates after contact with the insert, and returns back into the atmosphere, leaving no residue to contaminate the workpiece, chips, machine tool, or operator, and it also eliminates disposal costs. This is particularly helpful for porous powder metal parts, which often require subsequent part cleaning operations to remove coolant residue.

Now these two processes have become commercially available.

Words and Expressions

laser-assisted machining 激光辅助切削,激光辅助加工
cryogenic [ˌkraiəu'dʒenik] *a*. 低温学的
cryogenic machining 低温切削
laser beam 激光束
project onto 投射到……上
fiber optics 纤维光学,光导纤维,光纤
optical-beam 光束

chip 切屑
difficult-to-machine material 难加工材料
superalloy [ˌsjuːpəˈæləi] ***n***. 高温合金，超耐热合金
turning [ˈtəːniŋ] ***n***. 车削
cutting zone 切削区
physical problem 实际问题
insert 刀片
cubic boron nitride（**CBN**）立方氮化硼
polycrystalline cubic boron nitride（**PCBN**）聚晶立方氮化硼
indexable turning tool 可转位车刀
hard-turning 硬态车削（通常指对淬硬钢进行车削加工）
hard turning dry 硬态干式车削（在不使用任何切削液的情况下进行硬态车削）
fracture toughness 断裂韧度
water-based coolant 水基切削液，水基冷却液
oil-based coolant 油基切削液，油基冷却液
flank wear 后刀面磨损
alumina ceramic insert 氧化铝陶瓷刀片
finishing operation 精加工
lathe [leið] ***n***. 车床；***v***. 用车床加工
cylinder [ˈsilində] ***n***. 圆柱形容器，圆柱形物体，汽缸，柱面
turret [ˈtʌrit] ***n***. 转塔刀架
coupling [ˈkʌpliŋ] ***n***. 联轴器，管接头
nozzle [ˈnɔzl] ***n***. 喷嘴
residue [ˈrezidjuː] ***n***. 残留物
disposal cost 处理费用
powder metal 粉末金属（通过粉末方式得到的金属）

41 Effect of Reliability on Product Salability

Is reliability a saleable commodity? Yes and no. It all depends on the type of industry you are in. In the aerospace industry one can give an unqualified yes. Considering the technical complexity of their missions, the reliability achieved by the Apollo space missions is remarkable. One hundred per cent safe returns of the Apollo astronauts must be the reliability success story of the second half of the twentieth century. Did someone say why? Simply because reliability was written into the specification along with other performance parameters at the conceptual stage of the project. Tenders were therefore on the basis of achieving the standard of reliability written into the specification.

In the global market economy, the desire for profit provides the incentive to reach a high level of reliability. The individual firm or organization must assess the level of reliability necessary to enable the required sales targets to be achieved, along with the other engineering parameters of performance and operating characteristics. The production costs that must be maintained to enable the product to be put on the market at a competitive selling price will usually dictate the type of construction to be employed. Where this is novel, then the engineering development costs must allow for the spending associated with achieving the required level of reliability. Launching a new product which has an unsatisfactory reliability can make a complete nonsense of the best-laid marketing plans. The development work can, in many cases, be used as part of the marketing strategy, and can form an essential part of the launch program. It is also good insurance if the initial reliability is shown in the specification, for the customer is usually much less irate if he believes that the trouble has struck in spite of a sensible and visible program of work designed to check out the reliability of the product.

Examples of the combined development and marketing strategy come from the automotive industry, the aero-engine (see Fig. 41.1) industry, and the fan (see Fig. 41.2) industry. When a new model of car is announced it produces two opposite reactions in the potential customer's mind:

(1) It is new (and therefore "better" than the previous one).

(2) It is new (and therefore unproven and potentially full of teething troubles that will give unreliable operation).

Figure 41.1　Aero-engines

Figure 41.2　Fans

The strategy is to make the customer think it is new (and therefore better than the previous one and therefore desirable) and it has been well proven by test (and therefore should have few hidden faults which will require frequent visits to the service station). To do this it is vital to include in the publicity the more readily understood tests to prove the reliability of the new car, such as "this car has been driven for 100 000 miles before we let the public know of its existence," or "two hundred selected motorists have been given this car to test to make sure that our new car will stand up to the rigors of everyday business motoring." Both of these examples have been used in the past to tell the motoring public that extensive development work has been carried out on their behalf to arrive at a desirable and reliable product.

The strategy in the aero-engine industry is similar in that there must be a convincing reply to the airline or armed-forces customer who says that he can remember the teething troubles he had with the last new engine, and what has been done to make sure that he does not have to go through a similar period all over again? The simplest way to put over the development program is to show that with the previous new engine, 10 000 hours of engine running were carried out on the test bed before entering airline or military service, but with the new engine, 14 000 hours of running will be carried out on the test stand before service operation begins. In this sort of situation it is absolutely essential to use the very best market intelligence about what the customer really wants, and is really prepared to pay for.

In the fan industry the customer is concerned with price, performance, and reliability, and usually in that order. There is therefore likely to be less money in the kitty for elaborate testing, and every test must be meaningful to the engineer and convincing to the customer that he can install the product and forget it. Field testing is therefore very important for it is relatively cheap, but of course takes a fair amount of elapsed time. The customer here is particularly convinced by seeing installations, or details of installations, where the new product is running satisfactorily. Damp, steamy, or dirty installations are particularly desirable to introduce an element of "overstress" testing into the field service trials.

If the fan is to be installed in an agricultural situation, then it is important not to field test the fans with the most careful user that can be found, but rather with one who cleans down in a more haphazard fashion, hosing down the equipment when he should not. The hosing down may well show that a sealing arrangement needs to be incorporated on the shaft to stop water entering the motor and ruining the reliability.

Words and Expressions

salability [ˌseilə'biliti] *n.* 出售,销路,畅销
commodity [kə'mɔditi] *n.* 物品,商品,日用品,货物
unqualified ['ʌn'kwɔlifaid] *a.* 不合格的,无条件的,绝对的,彻底的
specification [ˌspesifi'keiʃən] *n.* 详细说明,尺寸规格,技术要求

tender [ˈtendə] *n.* 招标,投标

incentive [inˈsentiv] *a.* 刺激的,鼓励的;*n.* 刺激,鼓励,诱因,动机

launch [lɔːntʃ] *v.* 发射,创办,提出,开始

irate [aiˈreit] *a.* 发怒的,愤怒的

fan 通风机(是依靠输入的机械能,提高气体压力并排送气体的机械)

teethe [tiːð] *vi.* 出牙

teethe troubles 事情开始时的暂时困难

publicity [pʌbˈlisiti] *n.* 宣传材料,广告

motorist [ˈməutərist] *n.* 汽车驾驶人,开汽车的人

motoring [ˈməutəriŋ] *n.* 驾驶汽车;*a.* 汽车的

test bed 试验台

test stand 试验台

kitty [ˈkiti] *n.* 凑集的一笔钱,共同的资金

elapse [iˈlæps] *v.*,*n.* (时间)过去,消逝

overstress [əuvəˈstres] *n.*;*v.* 过载,超载,过度应力

haphazard [ˈhæpˈhæzəd] *n.* 偶然性,任意性;*a.* 杂乱的,任意的

hose down 用水龙带冲洗,用软管洗涤,用软管卸油

42 Reliability Requirements

Reliability requirements vary greatly depending on the type of equipment or system under consideration. They may sometimes be set by the designer or more broadly by the firm responsible for the design. In other situations the requirements may be imposed by the buyer of the product; and in some instances third outside parties, such as insurance underwriters or government agencies, may play a large role.

For any given product there are likely to be trade-offs between purchase prices and maintenance costs. The more effort put into making the product reliable, the higher will be its purchase price. At the same time, the more reliable a product is, the lower the costs for maintenance and repair are likely to be. We might be tempted to say that the solution is optimal when the total cost is minimized. In practice, however, the considerations that go into making such a trade-off vary greatly, depending on the nature of the product. The varying requirements may be understood more concretely by considering the factors that may come into play in the following three diverse examples: an air conditioner, an industrial robot, and an aircraft engine.

The reliability requirements for an air conditioner may depend to a great extent on consumer psychology. To what extent will the buyer be willing to pay a higher price in order to save later repair or replacement costs? The price is obvious, but how will the reliability be impressed on the general public: by information and advertising about the air conditioner's features, or by the willingness to offer a longer warranty than the competition? Moreover, what will the sales consequences be if the cost is too high because excessive reliability is built into the air conditioner? On the other hand, if the public buys the item at a lower price and then comes to resent the inconvenience and cost of breakdowns, will the firm's other products acquire a lingering reputation for poor design or shoddy workmanship? This, in turn, may depend significantly on the extent and efficiency of the service organization responsible for maintenance. These are complex issues. The decisions are most likely made by the manufacturer with the help of market surveys or other input to ascertain the public's preferences with regard to price and reliability.

The situation is quite different for an industrial robot (see Fig. 42.1) or other equipment that is designed primarily for sale to large organizations, such as manufacturing firms, public utilities, or government agencies. For if the equipment is expensive, the buyer is likely to have a staff of engineers or outside consultants who are able to assess the cost trade-offs and determine their impact on the buyer's operations. With the robot, for example, a trade-off must be made between purchase price and production lost through robot breakdown. The direct cost of repair may be small in comparison to the costs of lost production. In effect, the speed at which the buyer's maintenance organization is able to accomplish repairs will be a central consideration. Thus, for these types of products, design may center not only on reliability, but also on maintainability. Are they able to be repaired in a short period of time? Increasing the reliability at the expense of greatly increasing the repair time may in some situations cause greater losses in production.

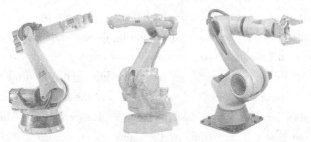

Figure 42.1　Industrial robots

For the aircraft engine the consequences of failure are generally so severe that a much higher standard of reliability is required, even though the cost of the engine is made much higher than it otherwise would be. Similarly, preventive maintenance to enhance reliability will be the primary consideration, rather than the time required to repair a failed engine. Avoiding failure is likely to be so important that it may not be left entirely in the hands of the design or manufacturing organization, or to the buyer, to set reliability specifications. Insurance underwriters and government agencies are likely to be much more closely involved. In the design more emphasis also will be placed on early warning of incipient failures so that the engine can be taken out of service before an accident occurs.

In setting reliability requirements and in designing to ensure that they are met, we should allocate reliability among major subsystems, and then down to lower levels of components and parts. For example, in designing an aircraft we might decide that the reliabilities of the propulsion, structural, control, and guidance systems should be approximately comparable. Then if the required reliability for the total system is to be equal to R, we would have $R = R_0^4$, where R_0 is the reliability requirement for each of the subsystems. Each subsystem reliability requirement would then be $R_0 = R^{1/4}$. Such reliability allocations are important particularly when the subsystems or components are to be designed or manufactured, or both, by different groups or by different corporations.

Thus far we have been interested only in the reliability, in the probability that the system will function properly. In reality not all failures can be treated alike, for there may be a wide range of consequences, depending on the mode of failure. In most system failures we may hope that the result is simply inconvenience and some economic loss. However, the designer must also consider carefully the possibility that a particular failure mode may endanger individuals or possibly the health or safety of a larger human population. The emphasis then becomes failure-oriented, for the primary goal is to prevent the system from causing an accident by the manner in which it fails.

We return to our diverse examples of an air conditioner, an industrial robot, and an aircraft engine to contrast safety analysis to more conventional reliability considerations. In most failures of an air conditioner its operation simply ceases. However, an electric short on an air conditioner may cause overheating and fire or even electrocute the user. The possibility of such accidents must be eliminated or reduced to a very small probability through careful design and safety analysis. If a faulty product causes such accidents, it may have to be withdrawn from the market, and the company may be brought to court in costly liability lawsuits.

The failure of the industrial robot is somewhat analogous, except that accidents are likely to be occupational risks to the workers involved, rather than risks to members of the general public. In contrast, when an aircraft engine fails there is little distinction between reliability and safety analysis; for if reliability is defined in terms of engine function, and failure has major safety implications because it will increase significantly the probability of a crash.

Words and Expressions

underwriter ['ʌndəraitə] *n.* 保险业者,保险商
trade-off 交换,协定,平衡,交易
concretely ['kɔnkri:tli] *ad.* 具体地
diverse [dai'və:s] *a.* 不同的,多种多种的
impression on 给……深刻印象
resent [ri'zent] *n.* 怨恨,生气,愤怒
lingering ['lingəriŋ] *a.* 延迟的,拖延的
shoddy ['ʃɔdi] *a.* 以次充好的,劣质的; *n.* 赝品
workmanship [wə:kmənʃip] *n.* 手艺,作工,工作质量
preference ['prefərəns] *n.* 偏爱,倾向,优先权
public utility 公用事业,公共事业机构
preventive maintenance 预防性维修,定期检修
in effect 实际上,事实上
liability [laiə'biliti] *n.* 责任,义务
analogous [ə'næləgəs] *a.* 类似的,相似的
incipient failure 早期故障,初期故障,将临故障
propulsion [prə'pʌlʃən] *n.* 推进,推进器
center on 集中,以……为中心
electrocute [i'lektrəkju:t] *v.* 触电致死
occupational [ˌɔkju'peiʃənəl] *a.* 职业的,占领的

43 Quality and Inspection

According to the American Society for Quality Control (ASQC), quality is the totality of features and characteristics of a product or service that bear on its ability to satisfy given needs. The definition implies that the needs of the customer must be identified first because satisfaction of those needs is the "bottom line" of achieving quality. Customer needs should then be transformed into product features and characteristics so that a design and the product specifications can be prepared.

In addition to a proper understanding of the term quality, it is important to understand the meaning of the terms quality management, quality assurance, and quality control.

Quality management is that aspect of the overall management function that determines and implements the quality policy. The responsibility for quality management belongs to senior management. This activity includes strategic planning, allocation of resources, and related quality program activities.

Quality assurance includes all the planned or systematic actions necessary to provide adequate confidence that a product or service will satisfy given needs. These actions are aimed at providing confidence that the quality system is working properly and include evaluating the adequacy of the designs and specifications or auditing the production operations for capability. Internal quality assurance aims at providing confidence to the management of a company, while external quality assurance provides assurance of product quality to those who buy from that company.

Quality control comprises the operational techniques and activities that sustain a quality of product or service so that the product will satisfy given needs. The quality control function is closest to the product in that various techniques and activities are used to monitor the process and to pursue the elimination of unsatisfactory sources of quality performance.

Many of the quality systems of the past were designed with the objective of sorting good products from bad products during the various processing steps. Those products judged to be bad had to be reworked to meet specifications. If they could not be reworked, they were scrapped. This type of system is known as a "detection correction" system. With this system, problems were not found until the products were inspected or when they were used by the customer. Because of the inherent nature of human inspectors, the effectiveness of the sorting operations was often less than 90%. Quality systems that are preventive in nature are being widely implemented. These systems prevent problems from occurring in the first place by placing emphasis on proper planning and problem prevention in all phases of the product cycle.

The final word on how well a product fulfills needs and expectations is given by the customers and users of that product and is influenced by the offerings of competitors that may also be available to those customers and users. It is important to recognize that this final word is formed over the entire life of the product, not just when it was purchased.

Being aware of customers' needs and expectations is very important, as was previously discussed. In addition, focusing the attention of all employees in an enterprise on the customers and users and their needs will result in a more effective quality system. For example, group discussions on product designs and specifications should include specific discussion of the needs to be satisfied.

A basic commitment of management should be that quality improvement must be relentlessly pursued. Actions should be ingrained in the day-to-day workings of the company that recognize that quality is a moving target in today's marketplace driven by constantly rising customer expectations. Traditional efforts that set a quality level perceived to be right for a product and direct all efforts to only maintain that level will not be successful in the long haul. Rather, management must orient the organization so that once the so-called right quality level for a product has been attained, improvement efforts continue to achieve progressively higher quality levels.

To achieve the most effective improvement efforts, management should understand that quality and cost are complementary and not conflicting objectives. Traditionally, recommendations were made to management that a choice had to be made between quality and cost, because better quality inevitably would somehow cost more and make production difficult. Experience throughout the world has shown that this is not true. Good quality fundamentally leads to good resource utilization and consequently means good productivity and low quality costs. Also significant is that higher sales and market penetration result from products that are perceived by customers to have high quality and performance reliability during use.

Four basic categories of quality costs are described in the following:

1. Prevention—costs incurred in planning, implementing, and maintaining a quality system that will ensure conformance to quality requirements at economical levels. An example of prevention cost is training in the use of statistical process control.

2. Appraisal—costs incurred in determining the degree of conformance to quality requirements. An example of appraisal cost is inspection.

3. Internal failure—costs arising when products, components, and materials fail to meet quality requirements prior to transfer of ownership to the customer. An example of internal failure cost is scrap.

4. External failure—costs incurred when products fail to meet quality requirements after transfer of ownership to the customer. An example of external failure cost is warranty claims.

Although the level of quality control is determined in large part by probability theory and statistical calculations, it is very important that the data collection processes on which these procedures depend be appropriate and accurate. The best statistical procedure is worthless if fed faulty data, and like machining processes, inspection data collection is itself a process with practical limits of accuracy, precision, resolution, and repeatability.

All inspection and/or measurement processes can be defined in terms of their accuracy and repeatability, just as a manufacturing process is evaluated for accuracy and repeatability. Controlled experiments can be performed, and statistical measures of the results can be made to determine the performance of a method of inspection relative to the parts to be inspected. Suitability of one or another method can be judged on the basis of standard deviations and confidence levels that apply to each approach as used in a given inspection situation.

Words and Expressions

bottom line 底线,要点,关键之处
strategic [strə'ti:dʒik] *a*. 战略的,关键性的,对全局有重大意义的
sort [sɔ:t] *n*. 种类,类别; *v*. 分类,拣选
rework ['ri:'wə:k] *v*. 重作,再加工,重新加工后利用
detection [di'tekʃən] *n*. 探测,发现
correction [kə'rekʃən] *n*. 改正,修正
commitment [kə'mitmənt] *n*. 委托,许诺,承担义务
relentlessly [ri'lentlisli] *ad*. 不屈不挠地,不懈地
ingrain [in'grein] *v*. 牢牢记在心中,使根深蒂固
long haul 长时间,持久,费时费力的工作
market penetration 市场覆盖率,市场渗透
trade off ['treidˌɔ:f] *n*. 折衷,权衡
irreversible [ˌiri'və:səbl] *a*. 不能撤回的,不能改变的
superficial [sju:pə'fiʃəl] *a*. 表面的,肤浅的,浅薄的
probability [prɔbə'biliti] *n*. 概率,可能性
resolution [rezə'lu:ʃən] *n*. 分辨率
repeatability 可重复性,再现性,反复性
suitability [sju:tə'biliti] *n*. 合适,适当,适宜性
standard deviation 标准偏差,标准差,均方差
confidence level 置信水平
controlled experiment 核对实验,对照实验,控制性实验

44 Coordinate Measuring Machines

A coordinate measuring machine (CMM) is a device for measuring the geometrical characteristics of an object. It is typically used to generate 3-D points from the surface of a part. It's digitizing a part in three dimensions. However, it is often used to make 2-D measurements such as measuring the center and radius of a circle in a plane, or even one-dimensional measurements such as determining the distance between two points. Typically, CMMs are configured to measure in Cartesian coordinates. There are also CMMs that measure in cylindrical or spherical coordinates. They can measure any part surface they can reach.

CMMs typically generate points in two ways: point-to-point mode, where the CMM touches the part and generates a single point of data every time contacting with the part; or scanning, where the CMM moves over a part, generating data as it moves. Scanning generates significantly more data than contacting.

The scanning, which includes laser scanning and white light scanning, is advancing very quickly. This method uses either laser beams or white light that are projected against the surface of the part. Many thousands of points can then be taken and used to not only check size and position, but to create a 3D image of the part as well. This "point-cloud data" can then be transferred to CAD software to create a 3D model of the part. These optical scanners often used on soft or delicate parts or to facilitate the "reverse engineering" process. Reverse engineering is the process of taking an existing part, measuring it to determine its size, and creating engineering drawings from these measurements. This is most often necessary in cases where engineering drawings may no longer exist or are unavailable for the particular part that needs replacement.

A typical stationary bridge-type CMM (Fig. 44.1) is composed of three axes, an X, Y and Z. These axes are orthogonal to each other in a typical three dimensional coordinate system. The CMM usually reads the input from the touch probe, as directed by the operator or program. The machine then uses the X, Y, Z coordinates of each of these points to determine size and position. Typical precision of a coordinate measuring machine is measured in micrometers.

Figure 44.1 Bridge-type CMM

A coordinate measuring machine (CMM) is also a device used in manufacturing and assembly processes to inspect a part or assembly against the design intent. By precisely recording the X, Y, and Z coordinates of the target, points are generated. These points are collected by using a probe that is positioned manually or automatically via computer control. In manual mode, the CMM is moved by the user. An automatic CMM is typically actuated by electric drives (using ball screws or linear motors, see Figs. 44.2 and 44.3).

As well as the traditional three axis machines, CMMs are also available in a variety of other forms. Portable CMMs are different from "traditional CMMs" in that they most commonly take the form of an articulated arm. As shown in Fig. 44.4, an articulated arm CMM looks very much like a six-degree-of-freedom robot. Articulated arm CMMs are lightweight (typically less than 10 kilograms) and can be carried and used nearly anywhere.

Figure 44.2 Ball screws

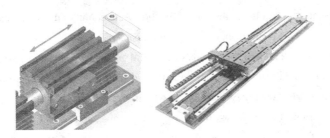

Figure 44.3 Linear motors

Figure 44.4 Articulated arm CMM

The accuracy of the traditional CMM is usually better than that of a mobile articulated arm CMM. But for many operations, the accuracy of articulated arm CMMs is sufficient for a variety of processes. The advantage of articulated arm CMMs is that they can reach areas that are not easy to access with traditional CMMs. Thus, if the accuracy of an articulated arm CMM is sufficient for a particular application, it should be seriously considered as an alternative.

To determine the level of accuracy needed from a CMM, the tightest tolerances to be measured should be considered first. As a rule of thumb, a CMM should have a level of accuracy 10 times greater than the tightest tolerance the shop needs to meet. In other words, if the tightest tolerance is 50 μm, the CMM should be accurate to 5 μm.

While the CMM hardware generates the coordinate data, the software bundled with the CMM (or in many instances sold separately) analyzes the data and presents the results to the user in a form that permits an understanding of part quality, and conformance to specified geometry.

The most important advancement in CMM technology over the past several years is that significant errors can be corrected mathematically via software. As a result, looser tolerances can be used on the system hardware, and the resulting errors (as long as they are highly repeatable) are eliminated in software. This results in lower manufacturing costs, while retaining or even improving the capabilities of the CMM.

New user-friendly software that allows the CMM and probe to be accurately, quickly, and easily calibrated has also made the CMM more accurate and easier to use.

A controlled environment is important for efficient CMM operation. CMMs can operate well on the shop floor if they are equipped with thermal compensation capabilities that correct for temperature changes from standard temperature (20℃). In any case, the CMM should be kept in a relatively clean environment and located in a space that is isolated from vibration.

Words and Expressions

coordinate measuring machine (CMM) 三坐标测量机
digitize ['didʒitaiz] *n.* (将资料)数字化
configure [kən'figə] *v.* (为特定设备或用途而进行的)设计, 配置, 使具一定形式
Cartesian coordinates 笛卡尔坐标, 直角坐标
point-to-point mode 点位方式, 从点到点的方式

scanning ['skæniŋ] *n*. 扫描
white light 白光
project ['prɔdʒekt] *v*. 投影,投射
point-cloud data 点云数据
delicate part 精密的零件,易损坏的零件
facilitate [fə'siliteit] *v*. (不以人作主语的)推动,帮助,使容易,促进
reverse engineering 反求工程,逆向工程
stationary bridge-type CMM 固定桥式三坐标测量机
orthogonal [ɔː'θɔgənl] *a*. 正交的,直角的,相互垂直的
assembly [ə'sembli] *n*. 装配,装配体,部件或机器
assembly process 装配过程
design intent 设计目的
probe [prəub] *n*. 测头,测量头;*v*. 探查,查明
electric drive 电气传动(生产过程中,以电动机作为原动机来带动生产机械,并按所给定的规律运动的电气设备)
ball screw 滚珠丝杠副(丝杠与旋合螺母之间以钢珠为滚动体的螺旋传动副。它可将旋转运动变为直线运动或将直线运动变为旋转运动)
linear motor 直线电机(一种将电能直接转换成直线运动机械能,而不需要任何中间转换机构的传动装置),也称线性电机
articulated [ɑː'tikjulitid] *a*. 铰接的,有关节的
articulated arm CMM 关节臂测量机
lightweight ['laitweit] *n*. 重量轻的,不重的
servo ['sɜːvəu] *n*. 伺服系统,伺服机构
bundled with 捆绑
conformance [kɔn'fɔːməns] *n*. 顺应,一致,符合
looser tolerance 比较大的公差
user-friendly software 用户友好的软件
calibrate ['kælibreit] *v*. 校准,标定,校验
controlled environment 受控环境
thermal compensation 热补偿

45 Computers in Manufacturing

The computer is bringing manufacturing into the Information Age. This new tool, long a familiar one in business and management operations, is moving into the factory, and its advent is changing manufacturing as certainly as the steam engine changed it more than 200 years ago.

The basic metalworking processes are not likely to change fundamentally, but their organization and control definitely will.

In one respect, manufacturing could be said to be coming full circle. The first manufacturing was a cottage industry: the designer was also the manufacturer, conceiving and fabricating products one at a time. Eventually, the concept of the interchangeability of parts was developed, production was separated into specialized functions, and identical parts were produced thousands at a time.

Today, although the designer and manufacturer may not become one again, the functions are being drawn close in the movement toward an integrated manufacturing system.

It is perhaps ironic that, at a time when the market demands a high degree of product diversification, the manufacturing enterprises have to increase productivity and reduce costs. Customers are demanding high quality and diversified products for less money.

The computer is the key to meet these requirements. It is the only tool that can provide the quick reflexes, the flexibility and speed, to meet a diversified market. And it is the only tool that enables the detailed analysis and the accessibility of accurate data necessary for the integration of the manufacturing system.

It may well be that, in the future, the computer may be essential to a company's survival. Many of today's businesses will fade away to be replaced by more-productive combinations. Such more-productive combinations are superquality, superproductivity plants. The goal is to design and operate a plant that would produce 100% satisfactory parts with good productivity.

A sophisticated, competitive world is requiring that manufacturing begin to settle for more, to become itself sophisticated. To meet competition, for example, a company will have to meet the somewhat conflicting demands for greater product diversification, higher quality, improved productivity, and low prices.

The company that seeks to meet these demands will need a sophisticated tool, one that will allow it to respond quickly to customer needs while getting the most out of its manufacturing resources.

The computer is that tool.

Becoming a "superquality, superproductivity" plant requires the integration of an extremely complex system. This can be accomplished only when all elements of manufacturing—design, fabrication and assembly, quality assurance, management, materials handling—are computer integrated.

In product design, for example, interactive computer-aided-design (CAD) systems allow the drawing and analysis tasks to be performed in a fraction of the time previously required and with greater accuracy. And programs for prototype testing and evaluation further speed the design process.

In manufacturing planning, computer-aided process planning permits the selection, from thousands of possible sequences and schedules, of the optimum process.

On the shop floor, distributed intelligence in the form of microprocessors controls machines, runs automated loading and unloading equipment, and collects data on current shop conditions.

But such isolated revolutions are not enough. What is needed is a totally automated system, linked by common software from front door to back.

Essentially, computer integration provides widely and instantaneously available, accurate information, improving communication between departments, permitting tighter control, and generally enhancing the overall quality and efficiency of the entire system.

Improved communication can mean, for example, designs that are more producible. The NC programmer and the tool designer have a chance to influence the product designer, and vice versa.

Engineering changes, thus, can be reduced, and those that are required can be handled more efficiently. Not only does the computer permit them to be specified more quickly, but it also alerts subsequent users of the data to the fact that a change has been made.

The instantaneous updating of production-control data permits better planning and more-effective scheduling. Expensive equipment, therefore, is used more productively, and parts move more efficiently through production, reducing work-in-process costs.

Product quality, too, can be improved. Not only are more-accurate designs produced, for example, but the use of design data by the quality-assurance department helps eliminate errors due to misunderstandings.

People are enabled to do their jobs better. By eliminating tedious calculations and paperwork—not to mention time wasted searching for information—the computer not only allows workers to be more productive but also frees them to do what only human beings can do: think creatively.

Computer integration may also lure new people into manufacturing. People are attracted because they want to work in a modern, technologically sophisticated environment.

In manufacturing engineering, CAD/CAM decreases tool-design, NC-programming, and planning times while speeding the response rate, which will eventually permit in-house staff to perform work that is currently being contracted out.

Words and Expressions

cottage industry 家庭手工业
conceive [kən'siːv] *v.* 设想,想象,表现,想出
interchangeability ['intəˌtʃeindʒə'biliti] *n.* 可交换性,互换性,可替代性
ironic [ai'rɔnik] *a.* 讽刺的,令人啼笑皆非的
diversification [daiˌvəːsifi'keiʃən] *n.* 多样化,变化,不同,多种经营

reflex ['ri:fleks] *n*. 反射,映像,反应能力
accessibility [æk‚sesi'biliti] *n*. 可接近性,易维护性,可存取性
it may well be that 完全可能,很可能是
fade away 渐渐消失
settle for 满足于,接受
conflicting [kən'fliktiŋ] *a*. 不一致的,冲突的,矛盾的,不相容的
accomplish [ə'kɔmpliʃ] *v*. 完成,达到目的,实行
shop floor 车间,工厂里的生产区,生产区的工人
distributed intelligence 分布式智能
schedule ['skedju:l] *n*. 时间表,进度表,计划表,进程,预订计划
detail ['di:teil] *n*. 细节,详细;*v*. 详述,画细部图,细部设计
update [ʌp'deit] *v*. 使……现代化,适时修正,不断改进,革新
work-in-process 在制品(即在一个企业的生产过程中,正在进行加工、装配或待进一步加工、装配或待检查验收的制品)
tedious ['ti:djəs] *a*. 冗长的,乏味的,慢的
lure [ljuə] *n*.;*v*. 吸引,诱惑,吸引力
in-house [in'haus] *a*. 国内的,机构内部的,公司内部的,自身的,固有的
contract out 订合同把工作包出去

46 Computer-Aided Design and Manufacturing

Computer-aided design (CAD) involves the use of computers to create design drawings and product models. Computer-aided design is usually associated with interactive computer graphics (known as a CAD system). Computer-aided design systems are powerful tools and are used in the mechanical design and geometric modeling of products and components.

In CAD, the drawing board is replaced by electronic input and output devices. When using a CAD system, the designer can conceptualize the object to be designed more easily on the graphics screen and can consider alternative designs or modify a particular design quickly to meet the necessary design requirements or changes. The designer can then subject the design to a variety of engineering analyses and can identify potential problems (such as an excessive load or deflection). The speed and accuracy of such analyses far surpass what is available from traditional methods.

Drafting efficiency is significantly improved. When something is drawn once, it never has to be drawn again. It can be retrieved from a library, and can be duplicated, stretched, sized, and changed in many ways without having to be redrawn. Cut and paste techniques are used as labor-saving aids.

CAD makes possible multiview 2D drawings, and the drawings can be reproduced at different levels of reduction and enlargement. It gives the mechanical engineer the ability to magnify even the smallest of components to ascertain if assembled components fit properly. Parts with different characteristics, such as movable or stationary, can be assigned different colors on the display.

Designers have even more freedom with the advent of 3D modeling. They can create 3D parts and manipulate them in endless variations to achieve the desired results. Through finite element analysis, stresses can be applied to a computer model and the results graphically displayed, giving the designer quick feedback on any inherent problems in a design before the creation of a physical prototype.

In addition to the design's geometric and dimensional features, other information (such as a list of materials, specifications, and manufacturing instructions) is stored in the CAD database. Using such information, the designer can then analyze the economics of alternative designs.

Computer-aided manufacturing (CAM) involves the use of computers and computer technology to assist in all the phases of manufacturing a product, including process and production planning, machining, scheduling, management, and quality control. Computer-aided design and computer-aided manufacturing are often combined into CAD/CAM systems.

This combination allows the transfer of information from the design stage into the stage of planning for the manufacture of a product, without the need to reenter the data on part geometry manually. The database developed during CAD is stored; then it is processed further, by CAM, into the necessary data and instructions for operating and controlling production machinery, material-handling equipment, and automated testing and inspection for product quality.

In machining operations, an important feature of CAD/CAM is its capability to describe the tool path for various operations, such as NC turning, milling, and drilling. The instructions (programs) are computer generated, and they can be modified by the programmer to optimize the tool path. The engineer or technician can then display and visually check the tool path for possible tool collisions with fixtures or other interferences. The tool path can be modified at any time, to accommodate other part shapes to be machined.

Some typical applications of CAD/CAM are: (a) programming for NC, CNC, and industrial robots; (b) design of tools and fixtures and EDM electrodes; (c) quality control and inspection, for instance, coordinate-measuring machines programmed on a CAD/CAM workstation; (d) process planning and scheduling; and (e) plant layout.

The emergence of CAD/CAM has had a major impact on manufacturing, by standardizing product development and by reducing design effort, tryout, and prototype work; it has made possible significantly reduced costs and improved productivity. The two-engine Boeing 777 passenger airplane, for example, was designed completely by computer (paperless design). The plane is constructed directly from the CAD/CAM software developed (an enhanced CATIA system) and no prototypes or mockups were built, such as were required for previous models.

Words and Expressions

design drawing 设计图
associated with 与……有关系，与……相联系
interactive computer graphics 交互式计算机制图学
geometric modeling 几何建模，几何造型
drawing board 绘图板，也可写为 **drafting board**
conceptualize [kən'septjuəlaiz] $v.$ 构思形成概念，产生想法
graphics screen 图形屏幕
drafting ['drɑːftiŋ] $n.$ 绘图，制图
library 库，程序库
paste [peist] $n.;v.$ 粘贴
labor-saving 省工，节省劳动力
multiview 多视图，多视角
enlargement [in'lɑːdʒmənt] $n.$ 放大
manipulate [mə'nipjuleit] $v.$ 处理，使用，利用，操纵
inherent problem 内在问题，固有问题
physical prototype 实体模型，实物原型
list of materials 物料清单，材料清单
scheduling ['ʃedjuːliŋ] $n.$ 编制进度计划
tool path 刀具轨迹，刀具路径
electrode [i'lektrəud] $n.$ 电极
plant layout 工厂布置，设备布置
tryout ['traiaʊnt] $n.$ 试验，试用
CATIA = Computer Aided Three-dimensional Interactive Application 计算机辅助三维交互应用
mockup ['mɔkˌʌp] $n.$ 实物模型

47 Computer-Aided Process Planning

According to the *Tool & Manufacturing Engineers Handbook*, process planning is the systematic determination of the methods by which a product is to be manufactured economically and competitively. It essentially involves selection, calculation, and documentation. Processes, machines, tools, operations, and sequences must be selected. Such factors as feeds, speeds, tolerances, dimensions, and costs must be calculated. Finally, documents in the form of process sheets, operation sheets, and process routes must be prepared. Process planning is an intermediate stage between designing and manufacturing the product. But how well does it bridge design and manufacturing?

Most manufacturing engineers would agree that, if ten different planners were asked to develop a process plan for the same part, they would probably come up with ten different plans. Obviously, all these plans cannot reflect the most efficient manufacturing methods, and, in fact, there is no guarantee that any one of them will constitute the optimum method for manufacturing the part.

What may be even more disturbing is that a process plan developed for a part during a current manufacturing program may be quite different from the plan developed for the same or similar part during a previous manufacturing program and it may never be used again for the same or similar part. That represents a lot of wasted effort and produces a great many inconsistencies in routing, tooling, labor requirements, costing, and possibly even purchase requirements.

Of course, process plans should not necessarily remain static. As lot sizes change and new technology, equipment, and processes become available, the most effective way to manufacture a particular part also changes, and those changes should be reflected in current process plans released to the shop.

A planner must manage and retrieve a great deal of data and many documents, including established standards, machinability data, machine specifications, tooling inventories, stock availability, and existing process plans. This is primarily an information-handling job, and the computer is an ideal companion.

There is another advantage to using computers to help with process planning. Because the task involves many interrelated activities, determining the optimum plan requires many iterations. Since computers can readily perform vast numbers of comparisons, many more alternative plans can be explored than would be possible manually.

A third advantage in the use of computer-aided process planning is uniformity.

Several specific benefits can be expected from the adoption of computer-aided process-planning techniques:

1. Reduced clerical effort in preparation of instructions.
2. Fewer calculation errors due to human error.
3. Fewer oversights in logic or instructions because of the prompting capability available with interactive computer programs.
4. Immediate access to up-to-date information from a central database.
5. Consistent information, because every planner accesses the same database.
6. Faster response to changes requested by engineers of other operating departments.
7. Automatic use of the latest revision of a part drawing. The design data would typically be the finished part (see Fig. 47.1) design file from the CAD system.
8. More-detailed, more-uniform process-plan statements produced by word-processing techniques.
9. More-effective use of inventories of tools, gages, and fixtures and a concomitant reduction in the variety of those items.
10. Better communication with shop personnel because plans can be more

specifically tailored to a particular task and presented in unambiguous, proven language.

11. Better information for production planning, including cutter-life forecasting, materials-requirements planning, scheduling, and inventory control.

Figure 47.1　Finished parts

Most important for CIM, computer-aided process planning produces machine-readable data instead of handwritten plans. Such data can readily be transferred to other systems within the CIM hierarchy for use in planning.

There are basically two approaches to computer-aided process planning: variant and generative.

In the variant approach, a set of standard process plans is established for all the parts families that have been identified through group technology. The standard plans are stored in computer memory and retrieved for new parts according to their family identification. Again, GT helps to place the new part in an appropriate family. The standard plan is then edited to suit the specific requirements of a particular job.

In the generative approach, an attempt is made to synthesize each individual plan using appropriate algorithms that define the various technological decisions that must be made in the course of manufacturing. In a truly generative process-planning system, the sequence of operations, as well as all the manufacturing-process parameters, would be automatically established without reference to prior plans. In its ultimate realization, such an approach would be universally applicable: present any part to the system, and the computer produces the optimum process plan.

No such system exists, however. So called generative process-planning systems—and probably for the foreseeable future—are still specialized systems developed for a specific operation or a particular type of manufacturing process. The logic is based on a combination of past practice and basic technology.

Words and Expressions

process planning 工艺过程设计,工艺设计
process sheet 工艺过程卡,工艺卡
operation sheet 工序卡片
process route 工艺路线
planner 工艺人员,也可写成 process planner
process plan 工艺规程
guarantee [ˌgærən'tiː] *n.*; *v.* 保证,保证书,担保,承认
lot size 批量
inventory ['invəntri] *n.* 清单,报表,总量,库存量; *v.* 开清单,清点,登记
companion [kəm'pænjən] *n.* 同伴,成对的物件之一,伙伴
clerical ['klerikəl] *a.* 书写的,文书的,事物性的
oversight ['əuvəsait] *n.* 监督,疏忽,忽略,误差
prompt [prɔmpt] *a.* 立即的,敏捷的,果断的; *v.* 促使,提醒,提示
up-to-date [ˌʌptə'deit] *a.* 现代化的,最新的,尖端的,当今的
concomitant [kən'kɔmitənt] *a.* 伴随的,相伴的,随……而产生的
tailor...to... 使……适合[满足]……(的要求,需要,条件等)
unambiguous [ˌʌnæm'bigjuəs] *a.* 明确的,清楚的,单值的,无歧义的
hierarchy ['haiərɑːki] *n.* 体系,系统,层次,分级结构
variant ['vεəriənt] *a.* 不同的,各种各样的; *n.* 变型,派生,衍生,转化
variant approach 派生法
generative ['dʒenərətiv] *a.* 能生产的,有生产力的,再生的,创成的
group technology 成组技术
algorithm ['ælgəriðəm] *n.* 算法,规则系统
part family 零件族,零件组
generative approach 创成法

48 Process Planning

Process planning is the determination of how a particular part will be manufactured in the shop. It consists of breaking the manufacturing process for the part into small steps, and then determining the optimum method of achieving that manufacturing step. This includes determining the type of manufacturing process to use, as well as the machining parameters for that particular process. It also includes determining a sequence of operations for the individual manufacturing actions that attempts to optimize a combination measure of manufacturing time and cost.

The process planning task can be reduced to five basic steps. First, the required operations must be determined by examining the design data and employing basic machining data (such as that square parts cannot be made on lathes).

Second, the machines required for each operation must be determined. This selection depends on knowledge of machine factors, such as availability, machining rate, size, power, and torque.

Third, the required tools for each identified machine must be determined.

Fourth, the optimum cutting parameters for each selected tool must be determined. These parameters include feed rate, cutting speed, depth of cut, and so forth. This determination depends on design data, such as material and surface roughness specifications, and on tool-cutting behavior.

Finally, an optimal combination of these manufacturing processes must be determined. The optimum is the process plan that minimizes some measure of manufacturing time and cost. This provides a detailed plan for the economical manufacture of the part.

Traditionally process planning is performed manually by highly experienced process planners who possess an in-depth knowledge of the manufacturing processes involved and the capabilities of the shop floor facilities. Because of the experience factor involved in process planning and in the absence of standardization of the process, conventional process planning has largely been subjective. Moreover, this activity is highly labor intensive. Rather than carrying out an exhaustive analysis and arriving at optimum values which could be time consuming, process planners often tend to play safe by using conservative values and this situation would invariably lead to no-optimal utilization of the manufacturing facilities and longer lead times.

During the last several decades, there has been considerable interest in computer-aided process planning (CAPP) —automating the process planning function by means of computer systems. Shop people knowledgeable in manufacturing processes are gradually retiring. An alternative approach to process planning is needed, and CAPP systems provide this alternative. Computer-aided process planning systems are designed around either of two approaches: retrieval systems and generative systems.

Retrieval CAPP System. Retrieval CAPP systems, also known as variant CAPP systems, are based on group technology and parts classification and coding. In these systems, the computer makes a search through its database of a number of standard or partially completed process plans that have been previously developed by human planners. Retrieval CAPP systems operate as indicated in Fig. 48.1. The user begins by identifying the GT code of the part for which the process plan is to be determined. A search is made of the part family file to determine if a standard process plan exists for the given part code. If the file contains a process plan for the part, it is retrieved and displayed for the user. The standard process plan is examined to determine whether modifications are necessary. Although the new part has the same code number, minor differences in the processes might be required to make the part. The standard plan is edited accordingly. The capacity to alter an existing process plan is why retrieval CAPP systems are also called variant CAPP systems.

If the file does not contain a standard process plan for the given code number, the user may search the file for a similar code number for which a standard process plan exists. By editing the existing process plan, or by starting from scratch, the user develops the process plan for the new part. This becomes the standard process plan for the new part code number.

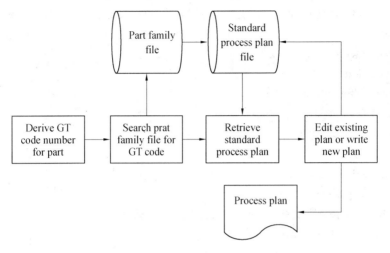

Figure 48.1 A retrieval CAPP system

Generative CAPP System. Generative CAPP systems are an alternative to retrieval systems. Rather than retrieving and editing existing plans from a database, a generative system creates the process plan using same logical procedures that would be followed by a traditional process planner in making that particular part. In a fully generative CAPP system, the process sequence is planned without human assistance and without predefined standard plans. However, the generative method is complex because it must contain comprehensive and detailed information of the part shape and dimensions; process capabilities; selection of the manufacturing methods, machinery, and tools; and the sequence of operations to be performed.

Designing a generative CAPP system is a problem in the field of expert systems, a branch of artificial intelligence. Expert systems are computer programs capable of solving complex problems that normally require a human who has years of education and experience.

Words and Expressions

manufacturing process 生产过程，制造过程，制造方法
step 工步
manufacturing step 工步
operation 工序
machining rate 加工速度
shop floor 车间，工场
feed rate 进给速度
in-depth 深入的，彻底的，全面的
exhaustive [ig′zɔ:stiv] *a*. 无遗漏的，彻底的，详尽的
process planner 工艺设计人员
labor intensive 劳动密集型的，要求或需要使用大量人力的
manufacturing facility 生产设备
manufacturing operation 制造过程
subjective [sʌb′dʒektiv] *a*. 主观的
invariably [in′veəriəb(ə)li] *ad*. 不变地，总是
alternative approach 替代方法，替代方式
retrieval CAPP system 检索式 CAPP 系统
variant CAPP system 派生式 CAPP 系统，变异型 CAPP 系统
parts classification and coding 零件分类和编码
standard process plan 典型工艺规程
GT code 成组编码
starting from scratch 从零开始
predefine [′pri:di′fain] *v*. 预先规定(或确定)
comprehensive and detailed information 全面和详细的信息
process capability 工序能力，加工能力
expert system 专家系统(一种模拟人类专家决策过程的计算机系统)
artificial intelligence 人工智能

49 Numerical Control

One of the most fundamental concepts in the area of advanced manufacturing technologies is numerical control (NC). Prior to the advent of NC, all machine tools were manually operated and controlled. Among the many limitations associated with manual control machine tools, perhaps none is more prominent than the limitation of operator skills. With manual control, the quality of the product is directly related to and limited to the skills of the operator. Numerical control represents the first major step away from human control of machine tools.

Numerical control means the control of machine tools and other manufacturing systems through the use of prerecorded, written symbolic instructions. Rather than operating a machine tool, an NC technician writes a program that issues operational instructions to the machine tool.

Numerical control was developed to overcome the limitation of human operators, and it has done so. Numerical control machines are more accurate than manually operated machines, they can produce parts more uniformly, they are faster, and the long-run tooling costs are lower. The development of NC led to the development of several other innovations in manufacturing technology:

1. Electrical discharge machining.
2. Laser cutting.
3. Electron beam welding.

Numerical control has also made machine tools more versatile than their manually operated predecessors. An NC machine tool can automatically produce a wide variety of parts, each involving an assortment of widely varied and complex machining processes. Numerical control has allowed manufacturers to undertake the production of products that would not have been feasible from an economic perspective using manually controlled machine tools and processes.

Like so many advanced technologies, NC was born in the laboratories of the Massachusetts Institute of Technology. The concept of NC was developed in the early 1950s with funding provided by the U.S. Air Force.

The APT (Automatically Programmed Tools) language was designed at the Servomechanism laboratory of MIT in 1956. This is a special programming language for NC that uses statements similar to English language to define the part geometry, describe the cutting tool configuration, and specify the necessary motions. The development of the APT language was a major step forward in the further development of NC technology. The original NC systems were vastly different from those used today. The machines had hardwired logic circuits. The instructional programs were written on punched paper tape (see Fig. 49.1), which was later to be replaced by magnetic plastic tape. A tape reader (see Fig. 49.2) was used to interpret the instructions written on the tape for the machine. Together, all of this represented a giant step forward in the control of machine tools. However, there were a number of problems with NC at this point in its development.

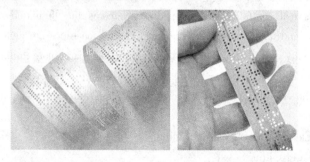

Figure 49.1　Punched paper tapes

A major problem was the fragility of the punched paper tape medium. It was common for the paper tape containing the programmed instructions to break or tear during a machining process. This problem was exacerbated by the fact that each successive time a part was produced on a machine tool, the paper tape carrying the programmed instructions had to be rerun through the reader. If it was necessary to produce 100 copies of a given part, it was also necessary to run the paper tape through the reader 100 separate times. Fragile paper tapes simply could not withstand the rigors of a shop floor environment and this kind of repeated use.

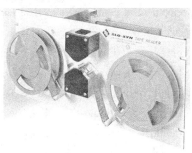

Figure 49.2 Photo-electric paper tape reader

This led to the development of a special magnetic plastic tape. Whereas the paper tape carried the programmed instructions as a series of holes punched in the tape, the plastic tape carried the instructions as a series of magnetic dots. The plastic tape was much stronger than the paper tape, which solved the problem of frequent tearing and breakage. However, it still left two other problems.

The most important of these was that it was difficult or impossible to change the instructions entered on the tape. To make even the most minor adjustments in a program of instructions, it was necessary to interrupt machining operations and make a new tape. It was also still necessary to run the tape through the reader as many times as there were parts to be produced. Fortunately, computer technology became a reality and soon solved the problems of NC associated with punched paper and plastic tape.

The development of a concept known as direct numerical control (DNC) solved the paper and plastic tape problems associated with numerical control by simply eliminating tape as the medium for carrying the programmed instructions. In direct numerical control, machine tools are tied, via a data transmission link, to a host computer. Programs for operating the machine tools are stored in the host computer and fed to the machine tool as needed via the data transmission linkage. Direct numerical control represented a major step forward over punched tape and plastic tape. However, it is subject to the same limitations as all technologies that depend on a host computer. When the host computer goes down, the machine tools also experience downtime. This problem led to the development of computer numerical control.

The development of the microprocessor allowed for the development of programmable logic controllers (PLCs) and microcomputers. These two technologies allowed for the development of computer numerical control (CNC). With CNC, each machine tool has a PLC (see Fig. 49.3) or a microcomputer that serves the same purpose. This allows programs to be input and stored at each individual machine tool. It also allows programs to be developed off-line and downloaded at the individual machine tool. CNC solved the problems associated with downtime of the host computer, but it introduced another problem known as data management. The same program might be loaded on ten different microcomputers with no communication among them. This problem is in the process of being solved by local area networks that connect microcomputers for better data management.

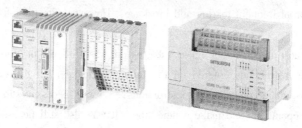

Figure 49.3 Programmable logic controllers

Words and Expressions

advent ['ædvənt] *n.* 到来,出现,来临
prerecord ['priːriˈkɔːd] *vt.* 预先录制
decode [diːˈkəud] *v.* 解译,译码,解码,译出指令
long-run *a.* 长期的,将来一定会发生的
tooling ['tuːliŋ] *n.* 工艺装备,刀具加工
discharge [disˈtʃɑːdʒ] *v.* 卸下,放出,放电

electrical discharge machining（EDM）电火花加工
welding［'weldiŋ］*n.*；*a.* 焊接(的),熔接(的),焊缝
predecessor［'priːdisesə］*n.* 前辈,被替代的原有事物,先驱
assortment［ə'sɔːtmənt］*n.* 种类,花色品种,分类,分级
feasible［'fiːzəbl］*a.* 可行的,做得到的,合理的,可用的
segment［'seɡmənt］*n.* 分割的部分,段,节,块
interpret［inˈtəːprit］*v.* 解释,说明,翻译,译码
fragility［frəˈdʒiliti］*n.* 脆弱,脆性,易碎性
punch［pʌntʃ］*n.*；*v.* 打孔,穿孔
photo-electric paper tape reader 光电纸带阅读器,光电纸带输入机
exacerbate［eksˈæsəːbeit］*vt.* 加重,使恶化,激怒
rerun［riːˈrʌn］*v.* 再开动,重新运转,重算
successive［səkˈsesiv］*a.* 连续的,接连的,逐次的,顺序的
via［'vaiə］*prep.* 经过,通过,借助于
programmable logic controller 可编程逻辑控制器
off-line 脱机,离线,指设备或装置不受中央处理机直接控制的情况

50 Computer Numerical Control

Today, computer numerical control (CNC) machine tools are widely used in manufacturing enterprises. Computer numerical control is the automated control of machine tools by a computer and computer program.

The CNC machines still perform essentially the same functions as manually operated machine tools, but movements of the machine tool are controlled electronically rather than by hand. CNC machine tools can produce the same parts over and over again with very little variation. They can run day and night, week after week, without getting tired. These are obvious advantages over manually operated machine tools, which need a great deal of human interaction in order to do anything.

A CNC machine tool differs from a manually operated machine tool only in respect to the specialized components that make up the CNC system. The CNC system can be further divided into three subsystems: control, drive, and feedback. All of these subsystems must work together to form a complete CNC system.

1. Control System

The centerpiece of the CNC system is the control. Technically the control is called the machine control unit (MCU), but the most common names used in recent years are controller, control unit, or just plain control. This is the computer that stores and reads the program and tells the other components what to do.

2. Drive System

The drive system is comprised of screws and motors that will finally turn the part program into motion. The first component of the typical drive system is a high-precision lead screw called a ball screw (Fig. 50.1). Eliminating backlash in a ball screw is very important for two reasons. First, high-precision positioning can not be achieved if the table is free to move slightly when it is supposed to be stationary. Second, material can be climb-cut safely if the backlash has been eliminated. Climb cutting is usually the most desirable method for machining on a CNC machine tool.

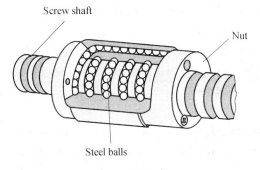

Figure 50.1　Ball screw

Drive motors are the second specialized component in the drive system. The turning of the motor will turn the ball screw to directly cause the machining table to move. Several types of electric motors are used on CNC control systems, and hydraulic motors (Fig. 50.2) are also occasionally used.

Figure 50.2　Hydraulic motors

The simplest type of electric motor used in CNC positioning systems is the stepper motor (sometimes called a stepping motor). A stepper motor rotates a fixed number of degrees when it receives an electrical pulse and then stops until another pulse is received. The stepping characteristic makes stepper motors easy to control.

It is more common to use servomotors in CNC systems today. Servomotors operate in a smooth, continuous motion—not like the discrete movements of the stepper motors. This smooth motion leads to highly desirable machining characteristics, but they are also difficult to control. Specialized hardware controls and feedback systems are needed to control and drive these motors. Alternating current (AC) servomotors are currently the standard choice for industrial CNC machine tools.

3. Feedback System

The function of a feedback system is to provide the control with information about the status of the motion control system, which is described in Fig. 50.3.

The control can compare the desired condition to the actual condition and make corrections. The most obvious information to be fed back to the control on a CNC machine tool is the position of the table and the velocity of the motors. Other information may also be fed back that is not directly related to motion control, such as the temperature of the motor and the load on the spindle—this information protects the machine from damage.

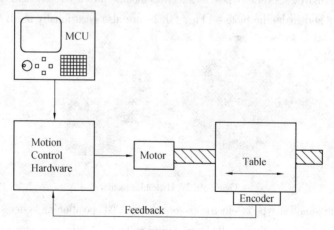

Figure 50.3 Typical motion control system of a CNC machine tool

There are two main types of control systems: open-loop and closed-loop. An open-loop system does not have any device to determine if the instructions were carried out. For example, in an open-loop system, the control could give instructions to turn the motor 10 revolutions. However, no information can come back to the control to tell it if it actually turned. All the control knows is that it delivered the instructions. Open-loop control is not used for critical systems, but it is a good choice for inexpensive motion control systems in which accuracy and reliability are not critical.

Closed-loop feedback uses external sensors to verify that certain conditions have been met. Of course, positioning and velocity feedback is of primary importance to an accurate CNC system. Feedback is the only way to ensure that the machine is behaving the way the control intended it to behave.

Words and Expressions

subsystem ['sʌbˌsistim] *n.* 子系统
centerpiece ['sentəpiːs] *n.* 主要特征；引人注目的东西
drive system 驱动系统，传动系统
backlash ['bæklæʃ] *n.* 反向间隙，回程误差
climb cutting 顺切，顺铣
screw shaft 丝杠，螺杆
drive motor 驱动电动机
ball screw 滚珠丝杠副
hydraulic motor 液压马达
stepper motor 步进电机
servomotor ['səːvəuˌməutə] *n.* 伺服电动机
feedback ['fiːdbæk] *n.* 反馈
alternating current 交流，交流电
open loop 开环
closed loop 闭环
encoder [in'kəudə] *n.* 编码器
critical system 关键系统，重要系统

51 Training Programmers

Skillful part-programmers are a vital requirement for effective utilization of NC machine tools. Upon their efforts depend the operational efficiency of those machines and the financial payback of the significant investment in the machines themselves, the plant's NC-support facilities, and the overhead costs involved.

Skillful NC part-programmers are scarce. This reflects not only the general shortage of experienced people in the metalworking industries but also the increasing demand for programmers as industry turns more and more to the use of numerically controlled machines to increase the capability, versatility, and productivity of manufacturing.

On an industry-wide basis, the obvious answer is to create new programmers by training them—and there are a number of sources for such training. But first, what qualifications should programmers have, and what must programming trainees learn?

According to the National Machine Tool Builders' Assn booklet "*Selecting an Appropriate NC programming Method*," the principal qualifications for manual programmers are as follows:

Manufacturing Experience Programmers must have a thorough understanding of the capabilities of the NC machine tools (see Fig. 51.1) being programmed, as well as an understanding of the basic capabilities of the other machines in the shop. They must have an extensive knowledge of, and sensitivity to, metalcutting principles and practices, cutting capabilities of the tools, and workholding fixtures and techniques. Programmers properly trained in these manufacturing engineering techniques can significantly reduce production costs.

Spatial Visualization Programmers must be able to visualize parts in three dimensions, the cutting motions of the machine, and potential interferences between the cutting tool, workpiece, fixture, or the machine itself.

Figure 51.1 NC machine tools

Mathematics A working knowledge of arithmetic, algebraic, trigonometric, and geometric operations is extremely important. A knowledge of higher mathematics, such as advanced algebra, calculus, etc, is not normally required.

Attention to Details It is essential that programmers be acutely observant and meticulously accurate individuals. Programming errors discovered during machine setup can be very expensive and time-consuming to correct.

"Manual programming," the booklet notes elsewhere, "requires the programmer to have more-detailed knowledge of the machine and control, machining practices, and methods of computation than does computer-aided programming. Computer-aided programming, on the other hand, requires a knowledge of the computer programming language and the computer system in order to process that language. In general, manual programming is more tedious and complex because of the detail involved. In a computer-aided programming system, this detail knowledge is embodied in the computer system (processor, postprocessor, etc)."

Experts in the NC and training fields typically agree on these qualifications and requirements—adding such subsidiary details as a knowledge of blueprint reading, machinability of different metals, use of shop measuring instruments, tolerancing methods, and safety practices.

Where should you look for candidates? First of all, in your own plant—

out on the shop floor. Edward F. Schloss, a Cincinnati Milacron sales vice president, puts it this way: "We've had excellent success with good lathe operators and good milling machine operators. They don't know it, but they've been programming most of their working lives, and they know basic shop math and trigonometry. You can teach them programming rather handily. Conversely, though, it's fairly hard to make NC part-programmers out of high-powered mathematicians. The path programming is easy. But what to do with it—the feeds, speeds, etc. —that may take even more-extensive training."

With more-powerful computer-assist programming, the need for metalcutting knowledge on the part of programmers is reduced. Through the use of this software, Cincinnati Milacron has been very successful in hiring new college graduates, including some with nontechnical degrees, and training them to be NC part-programmers. The trainees are given hands-on machine-tool experience in the plant before they are advanced to programming.

All suppliers of NC machine tools, of course, provide some sort of training in the programming of their products, and most offer formalized training programs, Milacron's sales department, for example, has 20 full-time customer-training instructors. The company's prerequisites for programmer training include the following:

"Participants must have knowledge of general machine-shop safety procedures and be able to read detail drawings, sectional views, and NC manuscripts.

" Knowledge of plane geometry, right-angle trigonometry, and fundamentals of tolerancing is required.

"Knowledge of NC manual part-programming, NC machine-tool setup and operating procedures, part processing, metalcutting technology, tooling, and fixturing is also needed."

Sending people with that kind of background to school will ensure that users of the NC machine will get the maximum benefit for their training dollar— the cost of a week of the trainee's time, travel, and living expenses, even though the training fee is waived with the basic purchase of the machine tool.

Words and Expressions

programmer ['prəugræmə] *n*. 程序设计员,程序编制员,订计划者
assn = association 协会,学会
booklet ['buklit] *n*. 小册子,目录单
NC machine tool 数控机床(也可写为 **CNC machine tool**)
workhold ['wəːkhəuld] *v*. 工件夹持
fixture ['fikstʃə] *n*. 工件夹具,固定物,定位器
visualization [ˌvizjuəlai'zeiʃən] *n*. 形象化,想象
algebraic [ˌældʒi'breiik] *a*. 代数的
trigonometric [ˌtrigənə'metrik] *a*. 三角学的,三角的
calculus ['kælkjuləs] *n*. 微积分,计算,演算
observant [əb'zəːvənt] *a*. 严格遵守……的,留心的,观察力敏锐的
meticulously [mi'tikjuləsli] *ad*. 过细地,细致地
setup ['setʌp] *n*. 装夹(装工件在机床上或夹具中定位、夹紧的过程),安装(工件或装配单元经一次装夹后所完成的那一部分工序),编排(在一个数控循环操作之前,确定用于控制和显示的一系列功能)
post processor 后置处理程序,后处理程序
subsidiary [səb'sidjəri] *a*. 辅助的,次要的,附属的;*n*. 附属机械
blueprint ['bluː'print] *n*.;*v*. (晒)蓝图,设计图,(订)计划
handily ['hændili] *ad*. 灵巧地,灵便地
hands-on *a*. 实习的,亲身试验的
prerequisite ['priː'rekwizit] *a*. 先决条件,必要的;*n*. 前提
waive [weiv] *v*. 放弃,推迟考虑,弃权,停止

52 NC Programming

The axial movements of CNC machine tools are guided by a computer, which reads a program and instructs several motors to move in the appropriate manner. The motors in turn cause the machine table to move and produce the machined part (see Fig. 52.1). Of course, the computer does not know what shape you want to cut until you write a program to describe the part.

Figure 52.1   Machined parts

The idea of controlling a machine tool electronically is rather simple. A programmer writes a set of instructions describing how he or she would like the machine to move and then feeds the program into the machine's computer. The computer reads the instructions and sends electrical signals to a motor, which then turns a screw to move the machine table. A sensor mounted on the table or on the motor sends positioning information back to the computer. Once the computer determines that the correct location has been reached, the next move will be executed.

Modern CNC machine tools (see also Fig. 51.1) can be programmed to perform machining operations with a language commonly referred to as G & M codes. G-codes and M-codes are simple instructions that the programmer will write to make the machine behave in certain way.

There are dozens of G-codes and M-codes used to perform everything from complex machining operations to ordinary tool changes. However, just a few codes will get most of the work done. The others are just enhancements to make programming more powerful and provide advanced machine functions.

The codes that are used to control and modify axis movements for machining and positioning operations are called preparatory codes. They are specified with the letter "G" and are usually referred to simply as G-codes. There are G-codes for many different machine operations, including positioning, straight line cutting, circular arc cutting, drilling operations, and many more.

G00 is the code used to perform rapid traverse. Rapid traverse is used for quickly positioning the tool in preparation for making a cut or moving the tool to a safe location for tool and part change. G00 can be used any place where the tool is not directly in contact with the workpiece. Rapid traverse is never used to perform a cut—the velocity is far too fast and is not constant.

G01 is the code used for linear interpolation. Interpolation is a term used to describe the process of passing through all points. In case of linear interpolation, the tool will travel through all points that make up a straight line in the Cartesian coordinate system. The G01 code moves the tool from one point to another at a specified velocity.

Circular arcs are another common feature in machining. To cut an arc, the codes G02 and G03 are used for circular interpolation. The two codes are identical in function, except that G02 will cut in a clockwise (CW) direction and G03 in a counterclockwise (CCW) direction.

In addition to G-codes, there are a number of codes used to control various machine functions ranging from tool change to coolant control. These codes are called miscellaneous codes or simply M-codes.

M-codes are called in a similar manner to G-codes. Many M-codes are stand alone instructions to carry out some function such as turning the spindle off. A few M-codes will also work in conjunction with other address to perform some function. For example, the code to turn on the spindle does not make much sense unless a spindle speed has been entered with an S address.

Tool changes are accomplished by calling the M06 code. After the tool change, it is a good idea to turn the spindle on before attempting to perform any metal cutting. This is accomplished with the code M03 and M04. M03 turns on the spindle in clockwise (CW). The spindle rpm must be specified in the same line or in a previous line. M04 is similar to M03, except it turns on the spindle in counterclockwise.

Virtually all end mills, drills, and similar tools you will encounter are right-handed and consequently use M03. However, there are situations when left-handed cutting tools are used and you will need to use M04.

Coolant can also be controlled with one of several M-codes. Coolants are often used in machining operations to extend the tool life, flush chips away, and improve the surface roughness.

M08 is the general code for turning on the coolant system. However, M07 may be used to differentiate between flood and mist coolant systems on machine tools equipped with both systems.

M09 is used to turn off the coolant. It is a good practice to turn the coolant off during tool changes to avoid spraying coolant into any critical areas. It is also a good idea to turn the coolant off when a pause is needed in the middle of the program to inspect the tool or workpiece.

There are many circumstances in which you will need to stop the program during execution in order to perform some task. For example, you may need to stop the program to clean chips out of the cutting zone.

Two codes will automatically cause program execution to stop: M00 and M01. The first, M00, is unconditional program stop. When the control reads the M00 code, it will stop all axis motion until the operator presses the cycle start button again. The M01 is effective only when the optional stop button on the control panel is depressed.

Words and Expressions

machine table 机床工作台
machined *a*. 经过机械加工的，已加工的
electrical signal 电信号
screw [skru:] *n*. 螺杆

G-code G 代码
positioning [pə'ziʃəniŋ] *n*. 位置控制,定位
preparatory code 准备功能代码
rapid traverse 快速移动
linear interpolation 直线插补
Cartesian coordinate 笛卡尔坐标,直角坐标
circular arc cut 圆弧切削
circular interpolation 圆弧插补
clockwise ['klɔkwaiz] *a*. 顺时针方向的; *adv*. 顺时针方向地
counterclockwise [ˌkauntə'klɔkwaiz] *a*. 逆时针方向的; *adv*. 逆时针方向地
stand alone 独立的
miscellaneous [misi'leiniəs] *a*. 混杂的,杂项的,各种各样的
miscellaneous code 辅助代码,辅助功能代码
call [kɔːl] *n*.; *v*. 称呼,调用,使计算机程序开始运行的动作
end mill 立铣刀
drill [dril] *n*. 钻头
right-handed 右旋的
left-handed 左旋的
differentiate [ˌdifə'renʃieit] *v*. 区别,区分
flood and mist coolant systems 液冷和雾冷系统
coolant ['kuːlənt] *n*. 冷冻剂,冷却液
unconditional ['ʌnkən'diʃənəl] *a*. 无条件的,无限制的
program stop 程序停止
optional stop 任选停止,可选择停止,计划停止
control panel 控制面板
depress [di'pres] *v*. 按下,压下,降低

53 Industrial Robots

There are a variety of definitions of the term industrial robot. Depending on the definition used, the number of industrial robot installations worldwide varies widely. Numerous single-purpose machines are used in manufacturing plants that might appear to be robots. These machines can only perform a single function and cannot be reprogrammed to perform a different function. Such single-purpose machines do not fit the definition for industrial robots that is becoming widely accepted.

An industrial robot is defined by the International Organization for Standardization (ISO) as an automatically controlled, reprogrammable, multipurpose manipulator, which may be either fixed in place or mobile for use in industrial automation applications.

There exist several other definitions too, given by other societies, e.g., by the Robot Institute of America (RIA), British Robot Association (BRA), and others. The definition developed by RIA is:

A robot is a reprogrammable multifunctional manipulator designed to move material, parts, tools, or specialized devices through variable programmed motions for the performance of a variety of tasks.

All definitions have two points in common. They all contain the words reprogrammable and multifunctional. It is these two characteristics that separate the true industrial robot from the various single-purpose machines used in modern manufacturing firms.

The term "reprogrammable" implies two things: The robot operates according to a written program, and this program can be rewritten to accommodate a variety of manufacturing tasks.

The term "multifunctional" means that the robot can, through reprogramming and the use of different end-effectors, perform a number of different manufacturing tasks. Definitions written around these two critical characteristics have become the accepted definitions among manufacturing professionals.

The first articulated arm came about in 1951 and was used by the U.S. Atomic Energy Commission. In 1954, the first industrial robot was designed by George C. Devol. It was an unsophisticated programmable materials handling machine.

The first commercially produced robot was developed in 1959. In 1962, the first industrial robot to be used on a production line was installed in the General Motors Corporation. It was used to lift red-hot door handles and other such car parts from die casting machines in an automobile factory in New Jersey, USA. Its most distinctive feature was a gripper that eliminated the need for man to touch car parts just made from molten metal. It had five degrees of freedom (DOF). This robot was produced by Unimation.

A major step forward in robot control occurred in 1973 with the development of the T^3 industrial robot by Cincinnati Milacron. The T^3 robot was the first commercially produced industrial robot controlled by a minicomputer. Figure 53.1 shows a T^3 robot with all the motions indicated, it is also called jointed-spherical robot.

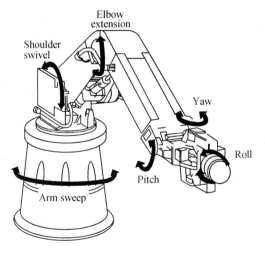

Figure 53.1　Jointed-spherical robot

Since then robotics has evolved in a multitude of directions, starting from using them in welding, painting, in assembly, machine tool loading and unloading, to inspection.

Over the last three decades automobile factories have become dominated by robots (see Fig. 53.2). A typical factory contains hundreds of industrial robots working on fully automated production lines. For example, on an automated production line, a vehicle chassis on a conveyor is welded, painted and finally assembled at a sequence of robot stations.

Figure 53.2　Robots used in automobile factories

Mass-produced printed circuit boards (PCBs) are almost exclusively assembled by pick-and-place robots, typically with SCARA manipulators, which pick tiny electronic components, and place them on to PCBs with great accuracy (see Figs. 53.3 and 53.4). Such robots can place tens of thousands of components per hour, far surpassing a human in speed, accuracy, and reliability.

A major reason for the growth in the use of industrial robots is their declining cost. Since 1970s, the rapid inflation of wages has tremendously increased the personnel costs of manufacturing firms. In order to survive, manufacturers were forced to consider any technological developments that could help improve productivity. It became imperative to produce better products at lower costs in order to be competitive in the global market economy. Other factors such as the need to find better ways of performing dangerous manufacturing tasks contributed to the development of industrial robots. However, the fundamental reason has always been, and is still, improved productivity.

 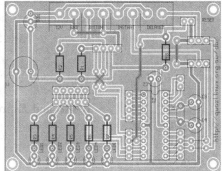

Figure 53.3　Printed circuit boards

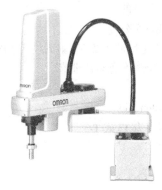 

Figure 53.4　SCARA manipulators

One of the principal advantages of robots is that they can be used in settings that are dangerous to humans. Welding and parting are examples of applications where robots can be used more safely than humans. Most industrial robots of today are designed to work in environments which are not safe and very difficult for human workers. For example, a robot can be designed to handle a very hot or very cold object that the human hand cannot handle safely.

Even though robots are closely associated with safety in the workplace, they can, in themselves, be dangerous. Robots and robot cells must be carefully designed and configured so that they do not endanger human workers and other machines. Robot workspaces should be accurately calculated and a danger zone surrounding the workspace clearly marked off. Barriers can be used to keep human workers out of a robot's workspace. Even with such precautions it is still a good idea to have an automatic shutdown system in situations where robots are used. Such a system should have the capacity to sense the need for an automatic shutdown of operations.

Words and Expressions

installation [ˌɪnstə'leiʃən] *n*. 整套装置,设备,结构,安装
multifunctional *a*. 多功能的
manipulator [mə'nipjuleitə] *n*. 机械手,机器人本体,操作机,操作臂
end effector 末端执行器,末端操作器
articulated [ɑː'tikjulitid] *a*. 关节式的,铰链的
General Motors Corporation (美国)通用汽车公司
die casting machine 压铸机(在压力作用下把熔融金属液压射到模具中冷却成型,开模后得到固体金属铸件的铸造机械)
gripper ['ɡripə] *n*. 手爪,夹爪,夹持器
Unimation 万能自动化公司(Universal Automation)
T^3 为 **The Tomorrow Tool** 的缩写,也可以写为 T3 或 T-3
jointed-spherical robot 关节式球面机器人
swivel ['swivl] *n*.;*v*. 旋转
yaw [jɔː] *n*. 侧摆,偏摆,偏转,摆动
pitch [pitʃ] *n*. 俯仰
roll [rəul] *n*. 侧滚,翻转,回转
chassis ['ʃæsi] *n*. 底盘
mass-produced 大批量生产的
printed circuit board 印刷电路板
pick-and-place robot 抓 – 放型机器人,抓放机器人,取 – 放型机器人
SCARA(**Selective Compliance Assembly Robot Arm**)选择顺应性装配机器手臂,平面关节型装配机器人
tens of thousands 成千上万,好几万
setting ['setiŋ] *n*. 位置,安装,环境
parting ['pɑːtiŋ] *a*. 分离的,离别的;*n*. 分离,切断

54　Robotics

An industrial robot is defined by ISO as an automatically controlled, reprogrammable, multipurpose manipulator programmable in three or more axes. The field of robotics may be more practically defined as the study, design and use of robot systems for manufacturing.

Robots may be classified by coordinate system, as follows:

1. Cartesian. In Cartesian coordinate robots, motion takes place along three linear perpendicular axes, that is, the arm can move up or down and in or out, in addition to a transverse motion perpendicular to the plane created by the previous two motions (Fig. 54.1a).

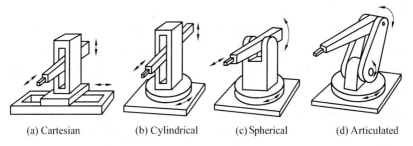

(a) Cartesian　　　(b) Cylindrical　　　(c) Spherical　　　(d) Articulated

Figure 54.1　Four types of industrial robots

2. Cylindrical. The motions of the arm of a cylindrical coordinate robot, like illustrated in Fig. 54.1b, are quite similar to those of the Cartesian coordinate robot, except that the motion of the base is rotary. In other words, the robot's arm can move up or down, in or out, and can swing around the vertical axis.

3. Spherical. The spherical coordinate robot, sometimes called polar coordinate robot, is shown in Fig. 54.1c. It has three axes of motion; two of them are rotary, and the third is linear.

4. Articulated. When a robot arm consists of links connected by revolute joints only, i.e. the prismatic joint in spherical type is also replaced by another revolute joint, the robot is called an articulated robot (Fig. 54.1d).

Robots may be attached permanently to the floor of a manufacturing plant, or they may move along overhead rails (gantry robots), or they may be equipped with wheels to move along the factory floor (mobile robots, such as AGVs used in factories for material handling purpose). A classification of robots by control method is described below.

1. Pick-and-Place Robot. The pick-and-place robot is programmed for a specific sequence of operations. Its movements are from point to point. These robots are simple and relatively inexpensive.

2. Playback Robot. An operator leads the playback robot and its end effector through the desired path; in other words, the operator teaches the robot by showing it what to do. The robot records the path and sequence of motions and can repeat them continually without any further action or guidance by the operator.

3. Numerically Controlled (NC) Robot. The numerically controlled robot is programmed and operated much like a NC machine. The robot is servo-controlled by digital data, and its sequence of movements can be changed with relative ease. As in NC machines, there are two basic types of controls: point-to-point, and continuous-path.

4. Intelligent Robot. The intelligent robot is capable of performing some of the functions and tasks carried out by human beings. It is equipped with a variety of sensors with visual and tactile capabilities. Much like humans, the robot can determine what action to make based on information acquired through its own sensors and decision making ability. Significant developments are taking place in intelligent robots so that they will

(1) Behave more and more like humans, performing tasks such as moving among a variety of machines and equipment on the shop floor and avoiding collisions;

(2) Recognize, select, and properly grip the correct raw material or workpiece;

(3) Transport the part to a machine for further processing or inspection;

(4) Assemble the components into subassemblies or a final product.

Major applications of industrial robots include the following:

1. Material handling consists of the loading, unloading, and transferring of workpieces in manufacturing facilities. These operations can be performed reliably and repeatedly with robots. Here are some examples: (a) casting operations, in which molten metal, raw materials, and parts in various stages of manufacture are handled automatically; (b) heat treating, in which parts are loaded and unloaded from furnaces and quench baths; (c) forming operations, in which parts are loaded and unloaded from presses and various other types of metalworking machinery.

2. Spot welding produce welds of good quality in the manufacture of automobile and truck bodies.

3. Operations such as deburring, grinding, and polishing can be done by using appropriate tools attached to the end effectors.

4. Spray painting (particularly of complex shapes) and cleaning operations are frequent applications because the motions for one piece repeat so accurately for the next.

5. Automated assembly is again very repetitive.

6. Inspection and gauging in various stages of manufacture make possible speeds much higher than those humans can achieve.

In addition to the technical factors, cost and benefit considerations are also significant aspects of robot selection and applications. The increasing availability and reliability, and the reduced costs, of sophisticated, intelligent robots are having a major economic impact on manufacturing operations, and such robots are gradually displacing human labor.

Words and Expressions

robotics [rəu'bɔtiks] *n.* 机器人学,机器人技术
Cartesian coordinate robot 直角坐标机器人,笛卡尔坐标机器人
in or out 伸缩
transverse motion 横向运动
cylindrical coordinate robot 柱面坐标机器人,圆柱坐标机器人
spherical coordinate robot 球面坐标机器人,球坐标机器人

polar coordinate robot 极坐标机器人
revolute joint 旋转关节
prismatic [priz'mætik] *a*. 棱柱的,棱柱形的
prismatic joint 移动关节
articulated robot 关节机器人
overhead rail 横梁,高架轨道
gantry robot 桁架式机器人,龙门式机器人
AGV (Automated Guided Vehicle) 自动导引车,无人搬运车
playback robot 示教再现式机器人
servo-controlled 伺服控制的
digital data 数字数据
point-to-point control 点位控制
continuous-path control 连续路径控制 连续轨迹控制
intelligent robot 智能机器人
decision making ability 决策能力
tactile ['tæktail] *a*. 触觉的,有触觉的
recognize ['rekəgnaiz] *v*. 辨认,感知
subassembly ['sʌbə'sembli] *n*. 部件,组件
manufacturing facility 生产设备
furnace 加热炉
quench bath 淬火槽
forming ['fɔːmiŋ] *n*. 成形,成形加工
press 冲床,压力机
spot welding 点焊
weld [weld] *v*. 焊接; *n*. 焊接,焊缝
deburring [di'bəːriŋ] *n*. 去毛刺,清除飞边
spray painting 喷漆
cleaning operation 清洗作业
gauging ['geidʒiŋ] *n*. 测量,测定,测试,计量
economic impact 经济影响
manufacturing operation 制造过程
displace [dis'pleis] *v*. 取代,置换

55 Basic Components of an Industrial Robot

To appreciate the functions of robot components and their capabilities, we might simultaneously observe the flexibility and capability of diverse movements of our arm, wrist, hand, and fingers in reaching for, and grabbing an object from a shelf, or in using a hand tool, or in operating a machine. Described next are the basic components of an industrial robot.

Manipulator. The manipulator is a mechanical unit that provides motions similar to those of a human arm and wrist. A manipulator is formed of links (their functionality being similar to the bones of the human body), and joints (also called "kinematic pairs") normally connected in series, as for the robots shown in Fig. 54.1. For a typical six degrees of freedom robot (Fig. 53. 1), the first three links and joints form the arm, and the last three joints make the wrist. The function of an arm is to place an object in certain location in the three-dimensional space, where the wrist orients it.

End Effector. End effectors are devices mounted on the end of the manipulator of a robot. It is equivalent to the human hand. Depending on the type of operation, conventional end effectors may be any of the following:

(1) Grippers, electromagnets, and vacuum cups, for material handling;

(2) Spray guns, for painting;

(3) Welding devices, for spot and arc welding;

(4) Power tools, such as electrical drills; and

(5) Measuring instruments.

End effectors are generally custom-made to meet special requirements. Mechanical grippers are the most commonly used and are equipped with two or more fingers. A two-fingered gripper (Fig. 55. 1) can only hold simple objects, whereas a multi-fingered gripper can perform more complex tasks. The selection of an appropriate end effector for a specific application depends on such factors as the payload, environment, reliability, and cost.

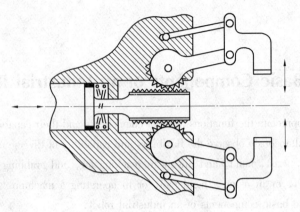

Figure 55.1 A simple gripper

Actuator. Actuators are like the "muscles" of a robot, they provide motion to the manipulator and the end effector. They are classified as pneumatic, hydraulic or electric, based on their principle of operation.

Pneumatic actuators utilize compressed air provided by a compressor and transform it into mechanical energy by means of pistons or air motors. Pneumatic actuators have few moving parts making them inherently reliable and reducing maintenance costs. It is the cheapest form of all actuators. But pneumatic actuators are not suitable for moving heavy loads under precise control due to the compressibility of air.

Hydraulic actuators utilize high-pressure fluid such as oil to transmit forces to the point of application desired. A hydraulic actuator is very similar in appearance to that of pneumatic actuator. Hydraulic actuators are designed to operate at much higher pressure (typically between 7 and 17 MPa). They are suitable for high power applications. Hydraulic robots are more capable of withstanding shock loads than electric robots.

Electric motors are the most popular actuator for manipulators. Direct current (DC) motors can achieve very high torque-to-volume ratios. They are also capable of high precision, fast acceleration, and high reliability. Although they don't have the power-to-weight ratio of hydraulic actuators or pneumatic actuators, their controllability makes them attractive for small and medium-sized manipulators.

Alternative current (AC) motors and stepper motors have been used infrequently in industrial robots. Difficulty of control of the former and low torque ability of the latter have limited their use.

Sensor. Sensors convert one form of signal to another. For example, the human eye converts light pattern into electrical signals. Sensors fall into one of the several types: vision, touch, position, force, velocity, acceleration etc. Some of them will be explained in Lesson 56.

Digital Controller. The digital controller is a special electronic device that has a CPU, memory, and sometimes a hard disk. In robotic systems, these components are kept inside a sealed box referred to as a controller. It is used for controlling the movements of the manipulator and the end effector. Since a computer has the same characteristics as those of a digital controller, it is also used as a robot controller.

Analog-to-Digital Converter. An analog-to-digital converter (abbreviated ADC) is an electronic device that converts analog signals to digital signals. This electronic device interfaces with the sensors and the robot's controller. Typically, an ADC converts an input analog voltage (or current) to a digital number proportional to the magnitude of the voltage or current. For example, the ADC converts the voltage signal developed due to the strain in a strain gage to a digital signal, so that the digital robot controller can processes this information.

Digital-to-Analog Converter (DAC). A DAC converts the digital signal from the robot controller into an analog signal to activate the actuators. In order to actually drive the actuators, e.g. a DC electric motor, the digital controller is also coupled with a DAC to convert its signal back to an equivalent analog signal, i.e., the electric voltage for the DC motors.

Amplifier. Generally, an amplifier is any device that changes, usually increases, the amplitude of a signal. Since the control commands from the digital controller converted to analogue signals by the DAC are very weak, they need amplification to really drive the electric motors of the robot manipulator.

Words and Expressions

reach for 伸手拿东西
functionality [ˌfʌŋkəʃəˈnæliti] *n*. 功能性
kinematic pair 运动副
in series 串联的
orient [ˈɔːriənt] *v*. 确定方向
gripper [ˈgripə] *n*. 手爪,夹持器
vacuum cup 真空吸盘
power tool 电动工具,动力工具
custom-made 定做的
payload [ˈpeiˌləud] *n*. 承载能力(carrying capacity),有效负载,有效载荷
actuator [ˈæktjueitə] *n*. 驱动器,执行机构
pneumatic [njuˈ(ː)ˈmætik] *a*. 空气的,气动的
pneumatic actuator 气动驱动器,气动执行机构
piston [ˈpistən] *n*. 活塞
air motor 气动马达
inherently [inˈhiərəntli] *adv*. 本能地,自然地,本质上地
compressibility [kəmˌpresiˈbiliti] *n*. 可压缩性
hydraulic actuator 液压驱动器
direct current (DC) 直流电
alternative current (AC) 交流电
controllability 可控性,可控制性
light pattern 光图像
analog-to-digital converter (ADC) 模数转换器,模拟–数字转换器
abbreviate [əˈbriːvieit] *v*. 节略,省略,缩写
interface [ˈintə(ː)ˌfeis] *n*. 界面,接口;*v*. 使连接,使接合:通过界面连接
strain gage 应变片,应变计
digital-to-analog converter 数模转换器
amplifier [ˈæmpliˌfaiə] *n*. 放大器

56 Robotic Sensors

Sensors in robots are like our eyes, nose, ears, mouth, and skin. Based on the function of human organs, for example the eyes or skin etc., terms like vision, tactile etc., have been used for robot sensors. Robots like humans, must gather extensive information about their environment in order to function effectively. They must pick-up an object and know it has been picked up. As the robot arm moves through the 3-dimensional space, it must avoid obstacles and approach items to be handled at a controlled speed. Some objects are heavy, others are fragile, and others are too hot to handle. These characteristics of objects and the environment must be recognized, and fed into the computer that controls a robot's movements. For example, to move the end effector of a robot along a desired trajectory and to exert a desired force on an object, the end effector and sensors must work in coordination with the robot controller. Robot sensors can be classified into two categories: internal sensors and external sensors.

Internal sensors, as the name suggests, are used to measure the internal state of a robot, i.e. its position, velocity, acceleration, etc. at a particular instant. Depending on the various quantities it measures, a sensor is termed as the position, velocity, acceleration, or force sensor.

Position sensors measure the position of each joint of a robot. There are several types of position sensors, e.g. encoder, LVDT, etc.

Encoder is an optical-electrical device that converts motion into a sequence of digital pulses. Encoders can be either absolute or incremental type. Further, each type may be again linear or rotary.

The linear variable differential transformer (LVDT) is one of the most used displacement transducers, particularly when high accuracy is needed. LVDT are made up of three coils, as shown in Fig. 56.1. AC supplied to the input coil at a specific voltage E_i generates a total output voltage across the secondary coils. The output voltage is a linear function of the displacement of a movable ferrous core inside the coils.

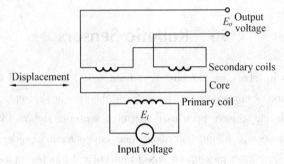

Figure 56.1 Schematic diagram of an LVDT

Electrical resistance strain gages are widely used to measure strains due to force or torque. Gages are made of electrical conductors, usually wire or foil, as shown in Fig. 56.2. They are glued on the surfaces where strains are to be measured. The strains cause changes in the resistance of the strain gages, which are measured by attaching them to the Wheatstone bridge circuit as one of the four resistances, $R_1 \ldots R_4$ of Fig. 56.3a. It is a cheap and accurate method of measuring strain. But care should be taken for the temperature changes. In order to enhance the output voltage and eliminate resistance changes due to the change in temperature, two strain gages are used, as shown in Fig. 56.3b, to measure the force at the end of the cantilever beam.

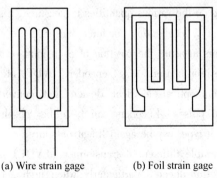

(a) Wire strain gage   (b) Foil strain gage

Figure 56.2 Strain gages

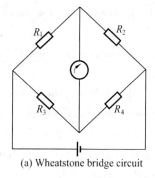 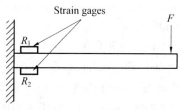

(a) Wheatstone bridge circuit   (b) Cantilever beam with strain gages

Figure 56.3   Strain measurement

A strain gage wrist sensor is shown in Fig. 56.4. It is milled from an aluminum tube 75 mm in diameter. Eight narrow elastic beams, four running vertically and four horizontally, have each been necked down at one point, so that the strain in each beam is concentrated at the other end of the beam. Strain gages cemented to each point of high strain measure the force component at that point.

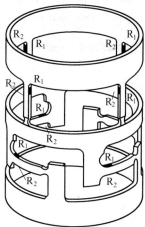

Figure 56.4   Strain gage wrist sensor

Foil strain gages, R_1 and R_2, are used to measure the strain. Two strain gages are mounted on each beam, so that the effects of temperature may be compensated for. This wrist sensor measures the three components of force and three components of torque in a Cartesian coordinate system.

External sensors are primarily used to learn more about the robot's environment. External sensors can be divided into two categories: contact type (e.g. limit switch) and non-contact type (e.g. proximity sensor and vision sensor).

A limit switch is constructed much as the ordinary light switch used at home and office. It has the same on/off characteristics. Limit switches are used in robots to detect the extreme positions of the motions, where the link reaching an extreme position switches off the corresponding actuator, thus safeguarding any possible damage to the mechanical structure of the robot arm.

A proximity sensor senses and indicates the presence of an object within a fixed space near the sensor. Proximity sensors are of two types, capacitive and inductive.

Vision systems are used with robots to let them look around and find the parts for picking and placing them at appropriate locations. Most current systems use cameras based on the charged coupled devices (CCDs) techniques. These cameras are smaller, last longer, and have less inherent image distortion than conventional robot cameras.

Words and Expressions

human organs 人体器官
fed into 被送入,被输入
trajectory [trəˈdʒekətəri] *n*. 轨迹
located at 位于,设在
desired trajectory 预定轨迹,期望轨迹
in coordination with 与……配合
internal sensor 内部传感器
external sensor 外部传感器
as the name suggests 顾名思义
at a particular instant 在某个具体瞬间
force sensor 力觉传感器,力传感器
encoder [inˈkəudə] *n*. 编码器,光电编码器
LVDT (Linear Variable Differential Transformer) 线性可变差动变压器,直线位移传感器

a sequence of 一系列的,一连串的

digital pulse 数字脉冲

absolute or incremental type 绝对式或增量式

linear or rotary 直线或旋转

most used 用得最多的

displacement transducer 位移传感器

coil [kɔil] *n*. 线圈

secondary coil 次级线圈

primary coil 初级线圈

ferrous core 铁芯

electrical resistance strain gage 电阻应变片

electrical conductor 导电体,电导体,电导线

attach to 连接到,加入,把……放在

Wheatstone bridge circuit 惠斯登电桥电路

care should be taken 特别要注意,值得注意

wire strain gage 电阻丝式应变片(也可以写作 **wire gage**)

foil strain gage 箔式应变片(也可以写作 **foil gage**)

wrist sensor 腕力传感器(也可以写作 **wrist force sensor**)

force component 分力

neck down 收缩,使……成颈状

concentrated at 集中在

limit switch 限位开关,行程开关

light switch 照明开关 电灯开关

proximity [prɔk'simiti] *n*. 接近

proximity sensor 接近觉传感器

capacitive [kə'pæsitiv] *a*. 电容的

inductive [in'dʌktiv] *a*. 电感的,感应的

current system 现行系统

charged coupled device (**CCD**) 电荷耦合元件,可以称为 **CCD** 图像传感器

image distortion 图像失真

57 Roles of Engineers in Manufacturing

Many engineers have as their function the designing of products that are to be brought into reality through the processing or fabrication of materials. They are a key factor in the selection of materials and manufacturing processes. A design engineer, better than any other person, should know what he or she wants a design to accomplish. He knows what assumptions he has made about service loads and requirements, what service environment the product must withstand, and what appearance he wants the final product to have. In order to meet these requirements he must select and specify the material(s) to be used. In most cases, in order to utilize the material and to enable the product to have the desired form, he knows that certain manufacturing processes will have to be employed. In many instances, the selection of a specific material may dictate what processing must be used. At the same time, when certain processes are to be used, the design may have to be modified in order for the process to be utilized effectively and economically. Certain dimensional tolerances can dictate the processing.

In any case, in the sequence of converting the design into reality, such decisions must be made by someone. In most instances they can be made most effectively at the design stage, by the designer if he has a reasonably adequate knowledge concerning materials and manufacturing processes. Otherwise, decisions may be made that will detract from the effectiveness of the product, or the product may be needlessly costly. It is thus apparent that design engineers are a vital factor in the manufacturing process, and it is indeed a blessing to the company if they can design for producibility—that is, for efficient production.

Manufacturing engineers select and coordinate specific processes and equipment to be used, or supervise and manage their use. Some design special tooling that is used so that standard machines can be utilized in producing specific products. These engineers must have a broad knowledge of machine and process capabilities and of materials, so that desired operations can be done effectively and efficiently without overloading or damaging machines and without adversely affecting the materials being processed. These manufacturing engineers also play an important role in manufacturing.

A relatively small group of engineers design the machines and equipment used in manufacturing. They obviously are design engineers and, relative to their products, they have the same concerns of the interrelationship of design, materials, and manufacturing processes. However, they have an even greater concern regarding the properties of the materials that their machines are going to process and the interreaction of the materials and the machines.

Still another group of engineers—the materials engineers—devote their major efforts toward developing new and better materials. They, too, must be concerned with how these materials can be processed and with the effects the processing will have on the properties of the materials.

Although their roles may be quite different, it is apparent that a large proportion of engineers must concern themselves with the interrelationship between materials and manufacturing processes.

Low-cost manufacture does not just happen. There is a close and interdependent relationship between the design of a product, selection of materials, selection of processes and equipment, and tooling selection and design. Each of these steps must be carefully considered, planned, and coordinated before manufacturing starts. This lead time, particularly for complicated products, may take months, even years, and the expenditure of large amount of money may be involved. Typically, the lead time for a completely new model of an automobile is about 2 years, for a modern aircraft it may be 4 years.

With the advent of computers and machines that can be controlled by computers, we are entering a new era of production planning. The integration of the design function and the manufacturing function through the computer is called CAD/CAM (computer aided design/computer aided manufacturing). The design is used to determine the manufacturing process planning and the programming information for the manufacturing processes themselves. Detail drawings can also be made from the central data base used for the design and manufacture, and programs can be generated to make the parts as needed. In addition, extensive computer aided testing and inspection (CATI) of the manufactured parts is taking place. There is no doubt that this trend will continue at ever-accelerating rates as computers become cheaper and smarter.

Words and Expressions

have as 把……作为
fabrication [fæbri'keiʃən] *n*. 制造,生产,加工,装配
assumption [ə'sʌmpʃən] *n*. 假设,设想,前提
procedure [prə'si:dʒə] *n*. 程序,工序,生产过程,方法,措施
sequence ['si:kwəns] *n*. 顺序,程序,工序,序列,连续
detract [di'trækt] *v*. 转移,毁损,贬低,减损
detract from 有损于,降低
interdependent [ˌintədi'pendənt] *a*. 相互依赖,相互影响,相互关联
interrelationship ['intəri'leiʃənʃip] *n*. 相互关系,相互联系,相互影响
advent ['ædvənt] *n*. 到来,出现,来临
with the advent of... 随着……的到来(出现)
integration [ˌinti'greiʃən] *n*. 积分,集成,综合,整体化
data base 数据库
in any case 无论如何,总之
needlessly ['ni:dlisli] *ad*. 不需要地,无用地,多余地
overload ['əuvələud] *v*.; *n*. (使)超载,超重,过负荷,使负担过重
complicated ['kɔmplikeitid] *a*. 错综复杂的,结构复杂的,麻烦的
expenditure [iks'penditʃə] *n*. 消费,支出,经费,开支,消费额
inspection [in'spektʃən] *n*. 检验,检查,观察,验收
ever-accelerating *a*. 不断加速的

58 Manufacturing Technology

Very simply, manufacturing is converting raw materials into usable products. A further refinement of this fundamental definition is that these products may be reproduced in quantity, at a quality which makes them functional, and at a cost which makes them competitive. There is a basic difference, therefore, between an artisan who designs and crafts a single chair, and a number of workers involved in building a quantity of identical chairs. This act of replication requires the organization of product design, materials, special tools and equipment, skilled people, and a management team into a production system.

The processes involved in converting materials into products can be classified as forming, cutting, and assembling. Ultimately, these are the only acts one can impose upon a substance to change it into something which people can use. Furthermore, these changes imposed upon materials are designed to produce modifications in the mass or bulk of the material, so that a subsequent change in geometry or shape occurs. These modifications can be described as mass or bulk conserving (forming), mass or bulk reducing (cutting), and mass or bulk increasing (assembling). Stated another way, the geometry or configuration of a material can be changed by moving material from one area to another, removing unnecessary material, or joining materials together.

Forming is a mass or bulk conserving process, where shape change is achieved by deformation. The bulk of the end product is equal, or almost equal, to the initial bulk of the workpiece or raw material.

The fundamental forming processes include bending, drawing, rolling, forging, extruding, and casting. In all but casting, the geometry change is effected by deformation. Metal can be squeezed, stretched, or twisted to produce various shapes. The casting of metals or ceramics involves the introduction of a fixed amount of material into a mold cavity. The material is conserved; only the precise amount needed goes into the finished product. Nothing is cut away to form the final part geometry. Forming becomes a significant method to produce pieces to near final shape and size at great economy, and with certain structural attributes. And, generally, it costs less.

Cutting is a mass or bulk reduction process, where shape change is achieved by removing material particles or chips until the workpiece has the desired geometry. Cutting therefore differs from hot or cold forming, where bulk deformation is employed to create a new shape, but where no gross part of the workpiece is removed.

Many shapes produced by cutting can be made by hot or cold forming or by casting. The greatest advantage of cutting is the ability to produce intricate shapes having extremely accurate dimensions. Material removal can be more precisely controlled than can plastic flow. Furthermore, the problems of shrinkage in hot-formed parts and the springback of cold-formed parts are eliminated. Because cutting does not rely on plastic flow to modify the shape of a workpiece, a cut part can have the same properties as the starting piece. For example, machined metal parts retain the benefits of prior cold work or heat treatment.

Cutting is usually more economical than forming for very complex parts, or when only a few pieces are to be produced. For most shapes, however, cutting is more expensive and must be used judiciously. In comparison with metal forming, for example, machining is slower, produces more waste material, and requires greater skill.

One of the greatest uses of cutting is to finish parts made by other processes. Many parts are produced by first making the rough shape by forming, and then completing the part by machining. Most turbine blades, for example, are milled from forged blanks. Some parts are produced almost entirely by forming, and then are machined only at areas where extremely accurate or complex shapes are required. For example, metal shafts for pumps or propellers are often hot- or cold-formed, with the bearing journals and threaded ends produced by machining.

Assembling involves joining together several components to create a final product. As such, it is a mass or bulk increasing process, because the total mass of that final product is the sum total of the masses of the individual parts of which it is made. Parts can be assembled by joining components with mechanical fasteners such as nuts and bolts (see Fig. 58.1), by fusing them together as weldments (see Fig. 58.2), or by adhering them with glues and solders. Some assemblies are meant to be permanent, while others are meant to be taken apart—disassembled—as required.

Figure 58.1   Nuts and bolts

Figure 58.2   Weldments

Any discussion of material processing leads to the ultimate conclusion that process involves selection and option. For example, holes may be generated in a piece of metal in a variety of ways. They may be drilled, milled, laser-cut, chemically-etched, or liquid jet-cut (see also Fig. 36.1). How do you cut the hole? The decision must be based on so many factors—the quality and accuracy of the hole, the reason for having a hole in the first place, the kind of metal being cut, the available equipment, the shape of the hole, numbers of parts, and speed, as well as others. The many people involved in the manufacture of the part must make the final decision and select the most appropriate process.

Words and Expressions

refinement [ri'fainmənt] *n*. 提纯,明确表达,改进的地方
functional ['fʌŋkʃənl] *a*.; *n*. 功能的,起作用的,实用的
in quantity 大量,大批
craft [krɑːft] *n*. 技能,工艺,行业; *v*. (用手工)精巧地制作
replication [ˌrepli'keiʃən] *n*. 重复,重现,仿作,复制过程
hot forming 热成形
ultimately ['ʌltimitli] *ad*. 毕竟,决定地,最后地,主要地,基本地
bulk [bʌlk] *n*. 容积,体积,大小,尺寸,基本部分; *a*. 体积的
mass or bulk conserving 质量或体积不变(守恒)
end product 最后产物,最终结果,成品
drawing ['drɔːiŋ] *n*. 拉拔,拉延,拉制,冲压成形
rolling ['rəuliŋ] *n*. 滚轧,滚压,压延; *a*. 滚压的,辊轧的
extruding [eks'truːdiŋ] *n*. 挤压成形,压制,模压
mold [məuld] = mould *n*. 模型,模具,模子
cavity ['kæviti] *n*. 空腔,孔穴,型腔,模腔
mold cavity 型模,阴模,铸模型腔
intricate ['intrikit] *a*. (错综)复杂的,交叉的
shrinkage ['ʃriŋkidʒ] *n*. 收缩,收缩率,减少
springback ['spriŋbæk] *n*. 回跳,回弹,弹回,弹性后效
starting ['stɑːtiŋ] *a*. 起初的,原来的,原始的; *n*. 起动,加快
cold work 冷加工,常温加工
judicious [dʒuː'diʃəs] *a*. 有见识的,合宜的,审慎的
as such 照这样,就点而论,因此
fuse [fjuːz] *v*. (使)熔化,熔合,联合
weldment ['weldmənt] *n*. 焊件,焊接件
adhere [əd'hiə] *v*. 粘着,附着,粘附
liquid jet-cut 液体射流切割
etch [etʃ] *v*.; *n*. 蚀刻,腐蚀(利用化学或电化学的方法对工件进行腐蚀或选择性腐蚀的工艺过程)

59 Mechanical Engineering in the Information Age

In the early 1980s, engineers thought that massive research would be needed to speed up product development. As it turns out, less research is actually needed because shortened product development cycles encourage engineers to use available technology. Developing a revolutionary technology for use in a new product is risky and prone to failure. Taking short steps is a safer and usually more successful approach to product development.

Shorter product development cycles are also beneficial in an engineering world in which both capital and labor are global. People who can design and manufacture various products can be found anywhere in the world, but containing a new idea is hard. Geographic distance is no longer a barrier to others finding out about your development six months into the process. If you've got a short development cycle, the situation is not catastrophic—as long as you maintain your lead. But if you're in the midst of a six-year development process and a competitor gets wind of your work, the project could be in more serious trouble.

The idea that engineers need to create a new design to solve every problem is quickly becoming obsolete. The first step in the modern design process is to browse the Internet or other information systems to see if someone else has already designed a transmission, or a heat exchanger that is close to what you need. Through these information systems, you may discover that someone already has manufacturing drawings, numerical control programs, and everything else required to manufacture your product. Engineers can then focus their professional competence on unsolved problems.

In tackling such problems, the availability of computers and access to computer networks dramatically enhance the capability of the engineering team and its productivity. These information age tools can give the team access to massive databases of material properties, standards, technologies, and successful designs. Such pretested designs can be downloaded for direct use or quickly modified to meet specific needs. Remote manufacturing, in which product instructions are sent out over a network, is also possible. You could end up with a virtual company where you don't have to see any hardware. When the product is completed, you can direct the manufacturer to drop-ship it to your customer. Periodic visits to the customer can be made to ensure that the product you designed is working according to the specifications. Although all of these developments won't apply equally to every company, the potential is there.

Custom design used to be left to small companies. Big companies sneered at it—they hated the idea of dealing with niche markets or small-volume custom solutions. "Here is my product," one of the big companies would say. "This is the best we can make it—you ought to like it. If you don't, there's smaller company down the street that will work on your problem."

Today, nearly every market is a niche market, because customers are selective. If you ignore the potential for tailoring your product to specific customers' needs, you will lose the major part of your market share—perhaps all of it. Since these niche markets are transient, your company needs to be in a position to respond to them quickly.

The emergence of niche markets and design on demand has altered the way engineers conduct research. Today, research is commonly directed toward solving particular problems. Although this situation is probably temporary, much uncommitted technology, developed at government expense or written off by major corporations, is available today at very low cost. Following modest modifications, such technology can often be used directly in product development, which allows many organizations to avoid the expense of an extensive research effort. Once the technology is free of major obstacles, the research effort can focus on overcoming the barriers to commercialization rather than on pursuing new and interesting, but undefined, alternatives.

When viewed in this perspective, engineering research must focus primarily on removing the barriers to rapid commercialization of known technologies. Much of this effort must address quality and reliability concerns, which are foremost in the minds of today's consumers. Clearly, a reputation for poor quality is synonymous with bad business. Everything possible—including thorough inspection at the end of the manufacturing line and automatic replacement of defective products—must be done to assure that the customer receives a properly functioning product.

Research has to focus on the cost benefit of factors such as reliability. As reliability increases, manufacturing costs and the final cost of the system will decrease. Having 30 percent junk at the end of a production line not only costs a fortune but also creates an opportunity for a competitor to take your idea and sell it to your customers.

Central to the process of improving reliability and lowering costs is the intensive and widespread use of design software, which allows engineers to speed up every stage of the design process. Shortening each stage, however, may not sufficiently reduce the time required for the entire process. Therefore, attention must also be devoted to concurrent engineering software with shared databases that can be accessed by all members of the design team.

As we move more fully into the Information Age, success will require that the engineer possess some unique knowledge of and experience in both the development and the management of technology. Success will require broad knowledge and skills as well as expertise in some key technologies and disciplines; it will also require a keen awareness of the social and economic factors at work in the marketplace. Increasingly, in the future, routine problems will not justify heavy engineering expenditures, and engineers will be expected to work cooperatively in solving more challenging, more demanding problems in substantially less time. It offers great promise and excitement as more and more problem-solving capability is placed in the hands of the computerized and wired engineer. We have begun a new phase in the practice of engineering. Mechanical engineering is a great profession, and it will become even greater as we make the most of the opportunities offered by the Information Age.

Words and Expressions

turn out 结果是，原来是，证明是
prone [prəun] *a.* 有……倾向，易于……的(to)
geographic [dʒiə'græfik] *a.* 地理上的，地区性的
get wind of ... 获得……线索，听到……风声
obsolete ['ɔbsəli:t] *a.* 已废弃的，已过时的；*n.* 废弃
browse [brauz] *v.*；*n.* 浏览，翻阅
tackle ['tækl] *v.* (着手)处理，从事，对付，解决
information age 信息时代
pretest ['pri:test] *n.*；*v.* 事先试验，预先检验
end up with ... 以……为结束，最后得出
virtual company 虚拟公司
niche market 瞄准机会的市场(指专门瞄准机会，做因市场不大而别人不做的产品，从而获得丰厚的利润)，特殊需求的市场，利基市场
sneer [sniə] *n.*；*v.* 冷笑，嘲笑，讥笑(at)
tailor ... to ... 使……适合[满足]……(要求，需要，条件等)
transient ['trænziənt] *a.* 短暂的，无常的，瞬变的
confer [kən'fə:] *v.* 授予，给予，使具有(性能)
modest ['mɔdist] *a.* 合适的，适度的，适中的，有节制的
evolutionary [,i:və'lu:ʃnəri] *a.* 发展的，发达的，进化的
uncommitted [,ʌnkə'mitid] *a.* 自由的，不受约束的，不负义务的，独立的
write off 抹去，勾销，消帐
undefined [,ʌndi'faind] *a.* 未规定的，不明确的，模糊的
junk [dʒʌŋk] *n.* 碎片，废物，废品；*v.* 丢掉，当作废物
fortune ['fɔ:tʃən] *n.* 运气，财富，财产
foremost ['fɔməust] *a.* 最初的，最主要的
concurrent engineering 并行工程(是对产品及其相关过程，包括制造过程和支持过程，进行并行、集成化处理的系统方法和综合技术。它要求产品开发人员从一开始就考虑到产品全生命周期内各阶段的因素，并强调各部门的协同工作)
make the most of 充分利用，极为重视

60　Mechanical Engineering and Mechanical Engineers

　　The field of mechanical engineering is concerned with machine components, the properties of forces, materials, energy, and motion, and with the application of the knowledge of those to devise new machines and products that improve society and people's lives.

　　Mechanical engineers research, develop, design, manufacture and test tools, machines, engines, and other mechanical devices. They work on power-producing machines such as electricity-producing generators, internal combustion engines, and gas turbines. They also develop power-using machines such as machine tools, robots, materials handling systems, and industrial production equipment. Mechanical engineers are known for working on a wide range of products and machines.

　　Mechanical engineers design equipment, it is produced by companies, and it is then sold to the public or to industrial customers in order to solve a problem and make an improvement. In the process of that business cycle, some aspect of the customer's life is improved and society as a whole benefits from the technical advances and additional opportunities that are offered by engineering research and development.

　　Mechanical engineering is not all about numbers, calculations, and gears. At its heart, the profession is driven by the desire to advance society through technology. Some of mechanical engineering's major achievements recognized in a survey conducted by ASME are the following:

　　1. **The automobile.** The development and commercialization of the automobile were judged by mechanical engineers as the profession's most significant achievement in the twentieth century. Two factors responsible for the growth of automotive technology have been high-power lightweight engines and efficient mass production processes.

Henry Ford pioneered the techniques of assembly-line mass production, which enabled consumers from across the economic spectrum to purchase and own automobiles. The automotive industry, in turn, has grown to become a key component of the world economy, and it has also created jobs in the machine tool and raw materials industries.

2. **Power generation.** Mechanical engineers make power-producing machines and convert energy from one form to another. Abundant and inexpensive energy is an important factor behind economic growth and prosperity, and the distribution of electrical energy has improved the lives and standard of living for billions of people across the globe. In the twentieth century, economies and societies changed significantly as electricity was produced and routed to homes, businesses, and factories.

3. **Agricultural mechanization.** The automation of farm equipment began with tractors and with the introduction of the combine, which simplified grain harvesting. Other advances have included improved high-capacity irrigation pumps, automated milking machines, and computer databases for the management of crops. With the advent of automation, more people have been able to take advantage of intellectual and employment opportunities in sectors of the economy other than agriculture. That migration of human resources away from agriculture has in turn helped to promote advances across a broad range of other professions and industries.

4. **Airplane.** Mechanical engineers have contributed to nearly every stage of aircraft research and development and have brought powered flight to its present advanced state. Commercial passenger aviation has created domestic and international travel opportunities for both business and recreational purposes. The advent of air travel has enabled geographically scattered families to visit one another frequently. Business transactions are conducted in face-to-face meetings between companies located on opposite sides of a country or halfway across the world.

5. **Integrated circuit mass production.** An amazing achievement of the electronics industry has been the miniaturization and mass production of integrated circuits, computer memory chips, and microprocessors.

An important process in producing semiconductor devices is lithography, in which complex circuit patterns are miniaturized and transferred from a pattern onto a silicon wafer. Transistors, capacitors, resistors, and wires are built as small as possible in order to fit more memory or computing power into a given space. Conducting wires, for instance, are as narrow as 120 nanometers in high-capacity dynamic random access memory chips. As a reference, a human hair is about 100 micrometers in diameter. Mechanical engineers work on the machines that manufacture those integrated circuits.

Words and Expressions

generator ['dʒenəreitə] *n.* 发电机，发生器
internal combustion engine 内燃机
gas turbine 燃气轮机
business cycle 商业周期
as a whole 整个来说，总体上
assembly-line 流水装配线
across the economic spectrum 不同经济地位，各个经济阶层
combine [kəm'bian] *n.* 康拜因，联合收割机
high capacity 大容量，高负载
irrigation pump 灌溉水泵
business transaction 商业交易，业务交易，交易往来
miniaturization ['minjətʃərai'zeiʃən] *n.* 小型化
integrated circuit 集成电路
memory chip 存储芯片
microprocessor [maikrəu'prəusesə] *n.* 微处理器
semiconductor ['semikən'dʌktə] *n.* 半导体
lithography [li'θɔgrəfi] *n.* 光刻，光刻法，平版印刷术
circuit pattern 电路图形，(印刷电路)电路图案
silicon wafer 硅片，晶圆
transistor [træn'zistə] *n.* 晶体管
capacitor [kə'pæsitə] *n.* 电容器(通常简称为电容)
resistor [ri'zistə] *n.* 电阻器(通常简称为电阻)
conducting wire 导线
nanometer ['nænə‚mitə] *n.* 纳米
dynamic random access memory 动态随机存储器

61 How to Write a Scientific Paper

Title In preparing a title for a paper, the author would do well to remember one salient fact: That title will be read by thousands of people. Perhaps few people, if any, will read the entire paper, but many people will read the title, either in the original journal or in one of the secondary (abstracting and indexing) services. Therefore, all words in the title should be chosen with great care, and their association with one another must be carefully managed.

The meaning and order of the words in the title are of importance to the potential reader who sees the title in the journal table of contents. But these considerations are equally important to all potential users of the literature, including those (probably a majority) who become aware of the paper via secondary sources. Thus, the title should be useful as a label accompanying the paper itself, and it also should be in a form suitable for the machine-indexing systems used by the *Engineering Index*, *Science Citation Index*, and others. Most of the indexing and abstracting services are geared to "key word" systems. Therefore, it is fundamentally important that the author provide the right "keys" to the paper when labeling it. That is, the terms in the title should be limited to those words that highlight the significant content of the paper in terms that are both understandable and retrievable.

Abstract An Abstract should be viewed as a mini-version of the paper. The Abstract should provide a brief summary of each of the main sections of the paper. A well-prepared abstract enables readers to identify the basic content of a document quickly and accurately, to determine its relevance to their interests, and thus to decide whether they need to read the document in its entirety. The Abstract should not exceed 250 words and should be designed to define clearly what is dealt with in the paper. Many people will read the Abstract, either in the original journal or in The *Engineering Index* or one of

3. Show how your results and interpretations agree (or contrast) with previously published work.
4. Don't be shy; discuss the theoretical implications of your work, as well as any possible practical applications.
5. State your conclusions as clearly as possible.
6. Summarize your evidence for each conclusion.

In showing the relationships among observed facts, you do not need to reach cosmic conclusions. Seldom will you be able to illuminate the whole truth; more often, the best you can do is shine a spotlight on one area of the truth. Your one area of truth can be supported by your data; if you extrapolate to a bigger picture than that shown by your data, you may appear foolish to the point that even your data-supported conclusions are cast into doubt.

Words and Expressions

table of contents 目录
relevance ['relivəns] *n.* 关联,关系,适用,中肯
provisional [prə'viʒənl] *a.* 暂定的,假定的,暂时的,临时的
rationale [ræʃiə'nɑːli] *n.* 基本原理,理论基础,原理的阐述
above all 尤其是,最重要的是,首先是
orient ['ɔːriənt] *v.* 定向,取向,正确地判断,(使)适应
obscure [əb'skjuə] *v.* 模糊的,不清楚的,难解的
injunction [in'dʒʌŋkʃən] *n.* 命令,指令
heed [hiːd] *v.*,*n.* 注意,留心
recapitulate [riːkə'pitjuleit] *v.* 扼要重述,概括,重现,再演
unsettled [ʌn'setld] *a.* 不稳定的,不安定的,未解决的,混乱的
correlation [kɔri'leiʃən] *n.* 关联,相关性,相互关系
fudge [fʌdʒ] *n.* 捏造,空话; *v.* 粗制乱造,捏造,推诿
implication [impli'keiʃən] *n.* 纠缠,隐含,意义,本质
cosmic ['kɔzmik] *a.* 宇宙的,全世界的,广大无边的
illuminate [i'ljuːmineit] *v.* 照亮,阐明,使明白,使显扬,使光辉灿烂
spotlight ['spɔtlait] *n.* 聚光灯,点光源,公众注意中心
extrapolate ['ekstrəpəleit] *n.* 推断,外推,外插

62 Technical Report Writing

Communication of your ideas and results is a very important aspect of engineering. Many engineering students picture themselves in professional practice spending most of their time doing calculations of a nature similar to those they have done as students. Fortunately, this is seldom the case. Actually, engineers spend the largest percentage of their time communicating with others, either orally or in writing.

When your design is done, it is usually necessary to present the results to your client, peers, or employer. The usual form of presentation is a formal technical report. Thus, it is very important for the engineering student to develop his or her communication skills. You may be the cleverest person in the world, but no one will know that if you cannot communicate your ideas clearly and concisely. In fact, if you cannot explain what you have done, you probably don't understand it yourself.

The following suggestions are presented as a guide to technical writing and an aid in avoiding some of the most common mistakes.

Title The title should be a meaningful description of what you have written.

Important Information Emphasize important information, beware of the common error of burying it under a mass of details.

Fact vs. Opinion Separate fact from opinion. It is important for the reader to know what your contributions are, what ideas you obtained from others (the references should indicate that), and which are opinions not substantiated by fact.

Tense The choice of the tense of verbs is often confusing to student writers. The following simple rules are usually employed by experienced writers:

Past tense. Use to describe work done in the laboratory or in general, to past evens, "Hardness readings were taken on all specimens."

Present tense. Use in reference to items and ideas in the report itself. "It is clear from the data in Figure 4 that strain energy is the driving force for recovery."

Future tense. Use in making prediction from the data that will be applicable in the future. "The data given in Table 2 indicate that in the next ten years, an increase in global temperature will cause sea levels to rise."

The following paragraphs provide the basics necessary for preparing each section of a technical report.

The Abstract—A Summary of the Entire Report. An Abstract must be a complete, concise distillation of the full report, and, as such, should always be written last. It should include a brief (one sentence) introduction to the subject, a statement of the problem, highlights of the results (quantitative, if possible), and major conclusions. It must stand alone without citing figures or tables. A concise, clear approach is essential, since most Abstracts are less than 250 words.

The Introduction—Why Did You Do What You Did? An Introduction generally identifies the subject of the report, provides the necessary background information including appropriate literature review, and, in general, provides the reader with a clear rationale for the work described. The Introduction does not contain results, and generally does not contain equations.

Analysis—What Does Theory Have to Say? An Analysis section describes a proposed theory or a descriptive model. It does not contain results, nor should extreme mathematical details be provided. Sufficient detail (mathematical or otherwise) should be provided for the reader to clearly understand the assumptions associated with a theory or model.

Experimental Procedure—What Did You Measure and How? The Experimental Procedure section is intended to describe how the experimental results were obtained and to describe any nonstandard types of apparatus or techniques that were employed. As a rule of thumb, provide sufficient details to allow the experiment to be conducted by someone else. If a list of equipment is included in the report, it should be a table in the body of the report, or should be placed in an appendix.

Results and Discussion—So What Did You Find?   Results of your work must be presented, as well as discussed, in this section of the report. Data must be interpreted to be useful to most readers. When presenting your results remember that even though you are usually writing to an experienced technical audience, what may be clear to you may not be obvious to the reader. Assuming too much knowledge can be a big mistake, so explain your results even if it seems unnecessary. If you can't figure them out, say so, "The mechanism is unclear and we are continuing to examine this phenomenon." Often the most important vehicles for the clear presentation of results are figures and tables. Column heads in tables should accurately describe the data that appear in that column. Each of the figures and tables should be numbered and have a descriptive title. Each table and figure must be explicitly and individually referenced and described in the text of the Results section.

Conclusions—What Do I Know Now?   The Conclusions section is where you should concisely restate your answer to the question: "What do I know now?" It is not a place to offer new facts, nor should it contain another rendition of experimental results or rationale. Conclusions should be clear and concise statements of the important findings of a particular study; most conclusions require some quantitative aspect to be useful.

Appendixes   Appendixes are the final elements in formal reports that contain supplemental information or information that is too detailed and technical to fit well into the body of the report or that some readers need and others do not. The recent trend in formal reports has been to place highly technical or statistical information in appendixes for those readers who are interested in such material.

Words and Expressions

picture ['piktʃə] *v*. 构想,想象
reference ['refrəns] *n*. 参考,参考文献,基准
substantiate [sʌbs'tænʃieit] *n*. 证实,证明
hardness ['hɑːdnis] *n*. 硬度

reading [ˈriːdiŋ] *n.* 读数，标度值，仪器指示值
driving force 驱动力
stand alone 独立的，独立存在的
rationale [ˌræʃəˈnɑːli] *n.* 基本原理
column head （表格中）各列内容的名称
as a rule of thumb 根据经验，一般说来
restate [ˈriːˈsteit] *v.* 重新叙述，重申
rendition [renˈdiʃən] *n.* 翻译，再现，解释
appendix [əˈpendiks] *n.* 附录，附属品
supplemental information 补充信息，附加信息

参考译文

第一课 力学基本概念

对运动、时间和作用力进行科学分析的分支称为力学。它由静力学和动力学两部分组成。静力学是对静止系统进行分析,即在其中不考虑时间这个因素,动力学则是对随时间而变化的系统进行分析。

力通过接触表面传到机器中的各构件上。例如,从齿轮传到轴或者从齿轮通过与其啮合的轮齿传到另一个齿轮,从V带传到带轮(见图1.1),或者从凸轮传到从动件(见图1.2)。由于很多原因,人们必须知道这些力的大小。例如,如果作用在一个滑动轴承上的力太大,它就会将油膜挤出,造成金属与金属的直接接触,产生过热和使轴承快速失效。如果齿轮相啮合的齿与齿之间的力过大,就会将油膜从齿间挤压出来。这会造成金属表层的碎裂和剥落,使噪音增大,运动不精确,直至报废。在力学研究中,我们主要关心力的大小、方向和作用点。

在力学中要用到的一些术语定义如下:

力 关于力的最早概念是由于我们需要推、提或拉各种物体而产生的。因此,力是物体之间的相互作用。力的直观概念包括作用点、方向和大小,这些被称为力的特性。

质量 是物体内所包含的物质的量。物体质量不受重力影响,因此它与物体的重量不同,但是与其成正比。尽管一块月球岩石在月亮上和地球上的重量不同,但它的物质含量是不变的。这不变的物质含量就称为岩石的质量。

惯性 是质量所具有的抵抗任何外力改变其本身运动状态的性质。

重量 是地球或其他天体对物体的作用力,它等于物体的质量与重力加速度的乘积。

质点 当一个物体的尺寸特别小,可以忽略不计时,该物体可以被称为质点。

刚体 刚体在受力后,其大小和形状都不会发生变化。实际上,所有的物体,不管是弹性体还是塑性体,在力的作用下都将发生变形。当物体的变形非常小时,为了简化计算,通常假设这个物体是刚体,也就是认为它没有发生变形。刚体是实际物体的理想化模型。

变形体 在分析由于外力的作用所引起物体内部的应力和应变时,不能采用刚体假设。这时,我们认为物体能够变形。这样的分析通常被称为弹性体分析,这时所用的假设为,在作用力的范围内,物体是弹性的。

牛顿运动定律 牛顿三定律为:

第一定律 如果作用在一个物体上的所有的力平衡,那么,这个物体将保持原来的静止或者匀速直线运动状态不变。

第二定律 如果作用在一个物体上的那些力不平衡,那么,这个物体将产生加速度。加速度的方向与合力的方向相同;加速度的大小与合力的大小成正比,与物体的质量成反比。

第三定律 相互作用的物体之间的作用力和反作用力大小相等,方向相反,作用在同一直线上。

力学涉及到两种类型的量:标量和矢量。标量是那些只有大小的量。在力学中标量的例子有时间,体积,密度,速率,能量和质量。另一方面,矢量既有大小又有方向。矢量的例子有位移,速度,加速度,力,力矩和动量。

第三课　力和力矩

当一些物体连接在一起形成一个组合或者系统时,任何两个相连接的物体之间的作用力和反作用力被称为约束力。这些力约束各个物体,使其处于特定的状态。从外部施加到这个物体系统的力被称为外力。

电力、磁力和重力是可以不需要通过接触而施加的力的实例。与我们有关的许多力则是必须通过直接的实际接触或机械接触才能产生。

力 F 是一个矢量。力的要素是它的大小、方向和作用点。力的方向包括力的作用线这个概念和它的指向。所以,沿着力的作用线,力的方向有正负之分。

作用在同一个刚体上的两条不重合平行线上的两个大小相等,方向相反的力不能合并成一个合力。作用在一个刚体上的这样两个力构成一个力偶。力偶臂是这两条作用线之间的垂直距离,力偶面是包含这两条作用线的平面。

力偶矩是一个垂直于力偶面的矢量 M,其指向根据右手法则确定。力偶矩的大小等于力偶臂与一个力的大小的乘积。

如果一个刚体满足下列条件,那么它处于静平衡状态:
(1)作用在它上面的所有外力的矢量和等于零。
(2)作用在它上面的所有外力对于任何一轴之矩的和等于零。
这两条可以用数学公式来表达:

$$\Sigma F = 0 \quad \Sigma M = 0$$

此外所使用的术语"刚体"可以是整台机器、一台机器中几个相互连接的零件、一个单独的零件或者零件的一部分。隔离体受力图是一个从机器中隔离出来的物体的草图或图,在图中标出所有作用在物体上的力和力矩(见图 3.1)。通常图中应该包括已知的力和力矩的大小与方向,以及一些其他有关的信息。

将这样得到的图称为"隔离体受力图"的原因是图中的零件或者物体的一部分已经从其余的机器零部件中隔离出来,其余的机器零部件对它的作用已经用力和力矩代替。对于一个完整的机器零件的隔离体受力图,图上所表示出的,作用在其上面的力和力矩是通过与其相邻或与其相连接的零件所施加的,是外力。对于一个零件的一部分的隔离体受力图,作用在切面上的力和力矩是由切掉部分所施加的,是内力。

绘制和提交简洁、清晰的隔离体受力图代表了工程交流的核心。这是因为它们是思考过程的一部分。整个思考过程包括了绘制在纸上和没有绘制在纸上的内容。这些图的绘制是将思考的结果与别人交流的唯一方式。不管出现的问题是多么简单,你应该养成绘制隔离体受力图的习惯。绘制隔离体受力图可以提高解决问题速度和大大减少产生错误的机会。

采用隔离体受力图的好处可以总结如下:
(1)能够使人们容易地把文字、思维和观念转换为物理模型。
(2)能够帮助人们观察和理解一个问题的各个方面。
(3)能够使数学关系变得容易理解或寻找。
(4)它们的应用易于记录解题的步骤,帮助作出有关简化的假设。
(5)解题所用的方法可以储存起来,供以后参考。
(6)能够帮助记忆,并且使你能够容易地向别人解释和展示你所做的工作。

在分析机器中的力时,我们几乎总是需要将机器分成许多单个的部件,绘制出表明每个部件受力情况的隔离体受力图。许多零部件将要通过运动副进行相互连接。

第五课 轴和联轴器

实际上,几乎所有的机器中都装有轴。轴最常见的形状是圆形,其截面可以是实心的,也可以是空心的(空心轴可以减轻重量)。有时也采用矩形轴,例如,一字螺丝刀的扁平形头部(见图 5.1)。

为了在传递扭矩时不发生过载,轴应该具有适当的抗扭强度。轴还应该具有足够的抗扭刚度,以确保同一个轴上的两个传动零件之间的相对转角不会过大。一般来说,当轴的长度等于其直径的 20 倍时,扭转角不应该超过 1 度。

轴安装在轴承中,通过齿轮、皮带轮、凸轮和离合器等零件传递动力。通过这些零件传来的力可能会使轴产生弯曲变形。因此,轴应该有足够的刚度以防止支承轴承受力过大。总而言之,在两个轴承支承之间,轴在每米长度上的弯曲变形不应该超过 0.5mm。

此外，轴还必须能够承受弯矩和扭矩的组合作用。因此，要考虑扭矩与弯矩的当量载荷。因为扭矩和弯矩会产生交变应力，在许用应力中也应该有一个考虑疲劳现象的安全系数。

直径小于 75mm 的轴可以采用含碳量大约为 0.4% 的冷轧钢，直径在 75～125mm 之间的轴可以采用冷轧钢或锻造毛坯。当直径大于 125mm 时，则要采用锻造毛坯，然后机械加工到所要求的尺寸。轻载时，广泛采用塑料轴。由于塑料是电的不良导体，在电器中用它作轴比较安全。

齿轮和皮带轮等零件通过键连接在轴上。在对键及轴上与之相应的键槽进行设计时，必须进行认真的计算。例如，轴上的键槽会引起应力集中，由于键槽的存在将使轴的横截面积减小，会进一步减弱轴的强度。

如图 5.2 所示，轴的直径发生变化时就会生成轴肩，其他机器零件可以通过靠在轴肩上进行定位。在轴肩处所产生的应力集中的大小取决于轴的两个直径的比值和轴肩处过渡圆角半径的大小。建议采用尽可能大的圆角半径，以减少应力集中。但是，有时齿轮、轴承或者其他零件会对过渡圆角半径产生影响。例如，轴承的内圈上有生产工厂加工出来的圆角半径（见图 11.1），但是，这个半径很小。如图 5.2a 所示，只有当过渡圆角半径更小时，才会保证轴承能够紧靠在轴肩上进行准确定位。如果一个零件中的孔有比较大的倒角，当这个零件通过紧靠在轴肩上进行定位时，过渡圆角半径就可以较大（如图 5.2b 所示），其相应的应力集中系数就会减小。

如果轴以临界速度转动，将会发生强烈的振动，可能会毁坏整台机器。知道这些临界转速的大小是很重要的，因为这样可以避开它。根据一般经验来说，工作速度与临界转速之间至少应该相差 20%。

轴的设计工作的另一个重要方面是轴与轴之间的直接联接方法。这是由刚性或者弹性联轴器来实现的。

联轴器是用来把两个相邻的轴端连接起来的装置。在机械结构中，联轴器被用来实现相邻的两根传轴之间的半永久性连接。在机器的正常使用期间内，这种连接一般不必拆开，在这种意义上，可以说联轴器的连接是永久性的。但是在紧急情况下，或者需要更换已磨损的零件时，可以先把联轴器拆开，然后再连接上。

联轴器有几种类型，它们的特性随其用途而定。如果制造工厂中或者船舶的螺旋桨需要一根特别长的轴，可以采用分段的方式将其制造出来，然后采用刚性联轴器将各段连接起来。

在把属于不同的设备（例如一个电动机和一个变速箱）的轴连接起来的时候，要把这些轴精确地对准是比较困难的，此时可以采用弹性联轴器。这种联轴器连接轴的方式可以把由于被连接的轴之间的轴线不重合所造成的有害影响减少到最低程度。弹性联轴器也允许被连接的轴在它们各自的载荷系统作用下产生偏斜或在轴线方向自由移动(浮动)而不至于相互干扰。弹性联轴器也可以用来减轻从一根轴传到另一根轴上的冲击载荷和振动的强度。

第七课　连接件和弹簧

连接件是将一个零件与另一个零件进行连接的零件。因此,几乎在所有的设计中都要用到连接件。人们对于任何产品的满意程度不仅仅取决于其组成部件,而且还取决于其连接方式。连接件为产品设计提供了以下特性:

(1)为检查和维修提供拆卸的方便;

(2)为由许多部件组成的组合式设计提供方便。采用组合式设计可以便于生产制造和运输。

连接件可以分为以下三类:

(1)可拆式。采用这种方式连接的零件很容易被拆开,而且不会对连接件造成损伤。例如普通的螺栓螺母连接。

(2)半永久式。采用此类方式连接的零件虽然能被拆开,但通常会对所用的连接件造成损伤,如开口销(图7.1)。

(3)永久式。采用这种连接件就表明所连接的零件不会被分开,例如铆接和焊接。

对于一个特定的应用,在选择连接件时应考虑以下几个方面:

(1)基本功能;

(2)外观;

(3)对于是采用大量的小型连接件还是采用少量的大型连接件的确定(螺栓可以作为一个例子);

(4)诸如载荷、振动和温度等工作条件;

(5)装拆频率;

(6)零件位置的可调性;

(7)被连接零件的材料种类;

(8)连接件失效或者松脱造成的后果。

通过任何一个复杂的产品,都能够认识到连接件在其中的重要性。以汽车为例,它是由数千个零件连接在一起而成为一辆整车的。一个连接件的失效或松脱可能会带来像车门嘎嘎响这类小麻烦,也可能造成像车轮脱落这种严重的后果。因此,在为一个特定的用途选择连接件时,应该考虑到上述各种可能性。

弹簧是一种能够在外载荷作用下,产生相当大的弹性变形的机械零件。描述变形与载荷成正比的虎克定律表明了弹簧的基本性能。然而,也有一些弹簧在其设计时所确定的载荷与变形之间的关系就是非线性的。弹簧的主要用途如下所述:

(1)控制机构运动。这类应用占弹簧用途的主要部分,如在离合器和制动器中起操纵力的作用。弹簧也被用来保持凸轮与从动件接触。

(2)缓冲和减振。这类应用包括汽车悬架系统弹簧和缓冲弹簧。

(3)储存能量。这类弹簧应用于钟表、摄影机和割草机中。

(4)测量力的大小。用来称体重的秤就是这类应用中最常见的一种。

弹簧主要可以分为压缩、拉伸和扭转弹簧这三种类型(如图7.2所示)。压缩弹簧和拉伸弹簧是最常用的弹簧。这些弹簧的变形是线性的。扭转弹簧的变形是角度而不是线性位移。板弹簧可以是简支梁型,也可以是悬臂梁型,它们的变形是线性的。

大部分弹簧是用钢制造的,但也有一些弹簧采用硅青铜、黄铜和铜铍合金制造。弹簧通常由专业生产弹簧的厂家制造。螺旋弹簧是最常用的弹簧,扭杆弹簧(图7.3)和板弹簧也有广泛的应用。对于螺旋弹簧,如果弹簧丝直径小于8 mm,通常采用冷拔钢丝或油回火钢丝经冷卷法制成。如果弹簧丝直径较大,采用热轧钢筋卷制弹簧。

选择弹簧,尤其是在遇到重载、高温、需要承受交变应力或者需要具有抗腐蚀性的时候;应该向弹簧制造厂家进行咨询。为了正确地选择弹簧,应该对弹簧的使用要求,包括空间限制进行全面的研究。有许多不同种类的专用弹簧可以满足一些特殊的要求或用途。

第九课 机械零件的强度

在设计任何机器或者结构时,所考虑的主要问题之一是其强度应该比它所承受的应力要大得多,以确保安全与可靠性。要保证机械零件在使用过程中不失效,就必须知道它们在某些时候为什么会失效的原因,然后,我们才能将应力与强度联系起来,以确保其安全。

设计任何机械零件的理想情况为,工程师可以利用大量的他所选用的这种材料的强度试验数据。这些试验应该采用与所设计的零件有着相同的热处理、表面粗糙度和尺寸的试件进行,而且试验应该在与零件使用过程中承受的载荷完全相同的情况下进行。这表明,如果零件将要承受弯曲载荷,那么就应该进行弯曲载荷的试验。如果零件将要承受弯曲和扭转的复合载荷,那么就应该进行弯曲和扭转复合载荷的试验。这些种类的试验可以提供非常有用和精确的数据。它们可以告诉工程师应该使用的安全系数和对于给定使用寿命时的可靠性。在设计过程中,只要能够获得这种数据,工程师就可以尽可能好地进行工程设计工作。

如果零件的失效可能会危害人的生命安全,或者零件有足够大的产量,则在设计前收集这样广泛的数据所花费的费用是值得的。例如,汽车和冰箱的零件的产量非常大,可以在生产之前对它们进行大量的试验,使其具有较高的可靠性。如果把进行这些试验的费用分摊到所生产的零件上的话,则每个零件摊到的费用是非常低的。

你可以对下列四种类型的设计作出评价:

(1)零件的失效可能会危害人的生命安全,或者零件的产量非常大,因此在设计时安排一个完善的试验程序会被认为是合理的。

(2)零件的产量足够大,可以进行适当的系列试验。

(3)零件的产量非常小,以至于进行试验根本不合算;或者要求很快地完成设计,以至于没有足够的时间进行试验。

(4)零件已经完成了设计、制造和试验,但其结果不能令人满意。这时候需要采用分析的方法来弄清楚不能令人满意的原因和应该如何进行改进。

我们将主要对后三种类型进行讨论。这就是说,设计人员通常只能利用那些公开发表的屈服强度、极限强度和延伸率等数据资料。人们期望着工程师在利用这些不是很多的数据资料的基础上,对静载荷与动载荷、两维应力状态与三维应力状态、高温与低温以及大零件与小零件进行设计!而设计中所能利用的数据通常是从简单的拉伸试验中得到的,其载荷是逐渐加上去的,有充分的时间产生应变。到目前为止,还必须利用这些数据来设计每分钟承受几千次复杂的动载荷的作用的零件,因此机械零件有时会失效是不足为奇的。

概括地说,设计人员所遇到的基本问题是,不论对于哪一种应力状态或者载荷情况,都只能利用通过简单拉伸试验所获得的数据并将其与零件的强度联系起来。

可能会有两种具有完全相同的强度和硬度值的金属,其中的一种由于其本身的延性而具有很好的承受超载荷的能力。延性是用材料断裂时的延伸率来量度的。通常将5%的延伸率定义为延性与脆性的分界线。断裂时延伸率小于5%的材料称为脆性材料,大于5%的称为延性材料。

材料的伸长量通常是在 50 mm 的计量长度上测量的。因为这并不是对实际应变量的测量,所以有时也采用另一种测量延性的方法。这个方法是在试件断裂后,测量其断裂处的横截面的面积。因此,延性可以表示为横截面的收缩率。

延性材料能够承受较大的过载荷这个特性是设计中的一个附加的安全因素。延性材料的重要性在于它是材料冷变形性能的衡量尺度。诸如弯曲和拉延这类金属加工都需要采用延性材料。

在选用抗磨损、抗侵蚀或者抗塑性变形的材料时,硬度通常是最主要的性能。

机构是构成许多机械设备的基本几何单元,这些机械设备包括自动包装机、夹具、机械式玩具、纺织机械等等。机构设计的目的通常是使一个刚体相对某一参考构件产生所需要的相对运动。机构的运动设计通常是设计一台完整的机器的第一步。当考虑力的作用时,应该考虑动力学、轴承载荷、应力、润滑等一系列问题。在所考虑的问题的范围扩大之后,机构设计就变成了机器设计。

作为机器的一个组成部分,机构的作用是在刚体之间传递或转换运动。常用的基本机构有以下三种:

齿轮机构 在这种机构中,各转轴之间的运动由相互啮合的齿轮来传递。齿轮通常用来传递角速度比为常值的运动,但是非圆齿轮可以用来传递角速度比为变数的运动。

凸轮机构 在这种机构中,输入件的等速连续运动被转换成输出件的不等速运动。输出的运动可以是轴的转动、滑块的移动、或者其他从动件的运动。这些运动都是使从动件与作为输入件的凸轮的轮廓的直接接触而产生的。凸轮的运动设计就是采用解析法或者图解法来确定凸轮的轮廓形状,使其能够带动从动件实现输出运动是输入运动的指定函数这一功能。

平面和空间连杆机构 这类机构也是用来使机构上某一点或者刚体实现机械运动的。连杆的基本作用有三种:

(1)刚体导向。刚体导向机构用来引导一个刚体,使其通过空间一系列预定的位置。

(2)实现轨迹。实现轨迹机构将引导刚体上的一个点,使其通过指定的空间轨迹上的一系列点。

(3)实现函数。这类机构所产生的输出运动是输入运动的指定函数。

为了强调各种机构之间的相同之处与不同之处,可以把它们按照几种不同的方式进行分类。一种分类方式是将机构分为平面、球面和空间三类。这三类机构有很多共同之处,然而,可以根据其构件的运动特点来确定分类准则。

在平面机构中,所有质点在空间所走过的轨迹都是平面曲线,所有这些平面曲线都位于相互平行的平面上,即所有点的轨迹都是平行于一个共同平面的平面曲线。这一特性使得平面机构上任意选定的一个点都可以按其真实尺寸和形状在一个平面图上表示出运动轨迹。平面四连杆机构、平板凸轮和它的从动件(参见图1.2)、曲柄滑块机构(见图13.1)是大家所熟悉的平面机构的例子。现在使用中的大多数机构是平面机构。

在球面机构中,当机构运动时,每一个构件上都有一个点是静止的,所有构件上的静止点都处于同一个位置,也就是每个点的轨迹都是球面曲线。所有各点运动时所在的球面都是同心的。因而,所有质点的运动都能用它们在以适当选取的点为中心的球面上的径向投影来完整地进行描述。虎克万向联轴器可能会是人们最熟悉的一个球面机构的例子。

从另一方面来说,在空间机构中质点的相对运动不受约束。运动的变换既不要求共面,也不要求同心。空间机构上许多质点的运动轨迹可能具有双重曲率。例如,任何含有螺旋副的连杆机构,由于其相对运动是螺旋线形的,因此是空间机构。

第十五课 摩擦学概论

摩擦学是一门研究在相对运动中相互作用表面的科学与技术。它来源于希腊语中的词 tribos,意思是摩擦。它研究工程表面的摩擦、润滑和磨损,目的是详细地理解表面间的相互作用,以便在实际应用中提出改进办法。摩擦学家的工作是跨学科的,包含有物理、化学、力学、热力学和材料科学等学科,并包括一个涉及有关表面间相对运动的机械设计、可靠性和工作性能的庞大复杂而且交织在一起的领域。

估计目前世界上大约有三分之一的能源是以各种摩擦形式消耗的。在今天的机械化社会中,这代表了潜在能量的一个惊人损失。研究摩擦学的目的是减少或消除在各种表面摩擦技术中不必要的浪费。

研究摩擦学的一个重要任务是按照我们的需要调节摩擦力的大小,例如可将其调到最小(如在机器中),或者最大(如作为防滑表面)。然而,必须着重指出,只有在对温度、滑动速度、润滑、表面粗糙度和材料性能等所有条件下的摩擦过程有了基本理解之后,这个目的才能实现。

从设计观点来看,对于一个实际应用的最重要判别标准,是在滑动界面上是干状态还是润滑状态起主导作用。在许多场合,例如在机器中,尽管可能存在着几种不同的润滑方式,只有一种状态(通常是润滑状态)起主导作用。在少数情况下,事先无法知道界面上究竟是干还是湿,显然这对于进行任何设计都是很困难的。这种现象最常见的例子是充气轮胎。在干摩擦的情况下,光滑的轮胎外表面在光滑的路面上可以获得最大的接触面积,使摩擦的粘附分量达到最大值。然而,这一组合在湿的情况下会产生非常低的摩擦系数,无法获得所期望的相对运动。在后一种情况下,如果在轮胎表面有合理的花纹和采用适当的路面结构,就能得到最佳的状态,尽管在干燥的气候下,这种组合会获得较低的摩擦系数。

润滑的方式可以划分为:流体动力润滑、边界润滑和弹性流体动力润滑。目前使用的各种轴承是完全流体动力润滑特性的最好例子,其滑动表面完全被界面间的润滑油膜隔开。边界润滑或者混合润滑是在相对运动表面之间既存在流体动力润滑,又有固体之间接触的一种混合状态。通常认为,在一个特定的产品设计中,当流体动力润滑失效时,才会出现这种混合状态。例如,一个滑动轴承在规定的载荷和速度下被设计成完全流体动力润滑的,但是当速度下降或者载荷增加时,就会造成轴颈和轴承表面一部分是固体接触,一部分是流体动力润滑这种状态。这种边界润滑是不稳定的,通常可以恢复到完全流体动力润滑,或者变成表面之间完全咬住。当薄的润滑油膜中的压力达到可以使润滑剂的边界面发生弹性变形时,则滑动界面上的润滑状态属于弹性流体动力润滑。

现在已经普遍认为在许多应用装置中存在着弹性流体动力接触状态,而过去却一直不确切地或者将其归类于流体动力润滑,或者将其归类于边界润滑状态,例如啮合齿轮的接触。固体润滑剂处于干状态和润滑状态之间,即尽管接触面通常是干的,但是固体润滑剂的存在使得它像在一开始就是湿润的一样。这是在特定的载荷和滑动条件下,固体润滑剂覆盖层表面上产生的一种物理——化学相互作用的结果,这样就产生了一种相当于润滑的效应。

第十七课 机械设计基础

机械设计是指机械装置和机械系统——机器、产品、结构、设备和仪器的设计。在大多数情况下,机械设计需要利用数学、材料科学和工程力学知识。

我们对整个设计过程感兴趣。它是怎样开始的?工程师是不是仅仅坐在铺着白纸的桌旁就可以开始设计了呢?当他记下一些设想后,下一步应该做些什么?什么因素会影响或者控制着应该作出的决定?最后,这一设计过程是如何结束的呢?

有时,虽然并不总是如此,工程师认识到一种需要并且决定对此做一些工作时,设计就开始了。认识到这种需要,并用语言将其清楚地叙述出来,常常是一种高度创造性的工作。因为这种需要可能只是一个模糊的不满,一种不舒服的感觉,或者是感觉到了某些东西是不正确的。

这种需要往往不是很明显的。例如,对食品包装机械进行改进的需要,可能是由于噪音过大、包装重量的变化、包装质量的微小的但是能够察觉得出来的变化等表现出来的。

叙述某种需要和确定随后要解决的问题之间有着明显的区别。要解决的问题是比较具体的。如果需要干净的空气,要解决的问题可能是降低发电厂烟囱的排尘量,或者是减少汽车排出的有害气体。

确定问题阶段应该制订设计对象所有的设计要求。这些设计要求包括输入量、输出量、设计对象的特性和它们所占据的空间尺寸以及对这些参量的所有制约因素。这些设计要求将规定生产成本、产量、预期寿命、工作温度和可靠性。

还存在着许多由于设计人员所处的特定环境或者由于问题本身的性质所产

生的隐含设计要求。某个工厂中可利用的制造工艺和设备会对设计人员的工作有所限制,因而成为隐含的设计要求的一部分。例如,一个小工厂中可能没有冷变形加工机械设备。因此,设计人员就必须选择这个工厂中能够进行的其他的金属加工方法。工人的技术水平和市场上的竞争情况也是隐含的设计要求的组成部分。

在确定了要解决的问题,并且形成了一组书面的和隐含的设计要求之后,设计工作的下一阶段是进行综合以获得最优的结果。因为只有通过对所设计的系统进行分析,才能确定其性能是否满足设计要求。因此,不进行分析和优化就不能进行综合。

设计工作是一个反复进行的过程。在这个过程中,我们要经历几个阶段,在对结果进行评价后,再返回到前面的阶段。因此,我们可以先综合系统中的几个零件,对它们进行分析和优化,然后再进行综合,看它们对系统的其他部分有什么影响。分析和优化都要求我们建立或者提出系统的抽象模型,以便对其进行数学分析。我们将这些模型称为数学模型。在建立数学模型时,我们希望能够找到一个可以很好地模拟实际物理系统的数学模型。

评价是整个设计过程中的一个重要阶段。评价是对一个成功的设计的最后检验,通常包括样机的实验室试验。在此阶段我们希望弄清楚设计能否真正满足所有的要求。它是否可靠?在与类似的产品的竞争中它能否获胜?制造和使用这种产品是否经济?它是否易于维护和调整?能否从它的销售或使用中获得利润?

与其他人就设计方案进行交流和沟通是设计过程的最后和关键阶段。毫无疑问,有许多伟大的设计、发明或创造之所以没有为人类所利用,就是因为创造者不善于或者不愿意向其他人介绍自己的成果。提出方案是一种说服别人的工作。当一个工程师向行政人员、管理人员或者其主管人员提出自己的新方案时,就是希望向他们说明或者证明自己的方案是比较好的。只有成功地完成这项工作,为得出这个方案所花费的大量时间和精力才不会被浪费掉。

人们基本上只有三种表达自己思想的方式,即文字材料、口头表述和绘图。因此,一个优秀的工程师除了掌握技术之外,还应该精通这三种表达方式。如果一个技术能力很强的人在上述三种表达方式中的某一种的能力较差,他就会遇到很大的困难。如果上述三种能力都较差,那将永远没有人知道他是一个多么能干的人!

一个有能力的工程师不应该害怕在提出自己的方案时遭到失败的可能性。事实上,偶然的失败肯定会发生的,因为每一个真正有创造性的设想似乎总是有失败或批评伴随着它。从一次失败中可以学到很多东西,只有不怕遭受失败的人制造任何产品的第一步工作都是设计。设计通常可以分为几个明确的阶段：(a)概念设计；(b)功能设计；(c)生产设计。在概念设计阶段，设计者着重考虑产品应该具有的功能。通常要设想和考虑几个方案，然后决定这种想法是否可行；如果可行，则应该对其中一个或几个方案作进一步的改进。在此阶段，关于材料选择唯一要考虑的问题是：是否有性能符合要求的材料可供选用；如果没有的话，是否有较大的把握在成本和时间都允许的限度内研制出一种新材料。

在功能设计或工程设计阶段，要做出一个切实可行的设计。在这个阶段要绘制出相当完整的图纸，选择并确定各种零件的材料。通常要制造出样机或者实物模型，并对其进行试验，评价产品的功能、可靠性、外观和维修保养性等。虽然这种试验可能会表明，在产品进入到生产阶段之前，应该更换某些材料，但是，绝对不能将这一点作为不认真选择材料的借口。应该结合产品的功能，认真仔细地考虑产品的外观、成本和可靠性。一个很有成就的公司在制造所有样机时，所选用的材料应该和其在生产中使用的材料相同，并尽可能使用同样的制造技术。这样做对公司是很有好处的。功能完备的样机如果不能根据预期的销售量经济地制造出来，或是样机与正式生产的装置在质量和可靠性方面有很大不同，则这种样机就没有多大的价值。设计工程师最好能在这一阶段全部完成材料的分析、选择和确定工作，而不是将其留到生产设计阶段去做。因为，在生产设计阶段材料的更换是由其他人进行的，这些人对产品的所有功能的了解可能不如设计工程师。

在生产设计阶段中，与材料有关的主要问题是应该把材料完全确定下来，使它们与现有的设备相适应，能够利用现有设备经济地进行加工，而且材料的数量能够比较容易地保证供应。

在制造过程中，不可避免地会出现对使用中的材料作一些更改的情况。经验表明，可以采用某些便宜材料作为替代品。然而，在大多数情况下，在进行生产以后改换材料要比在开始生产前改换材料所花费的代价要高。在生产设计阶段做好材料选择工作，可以避免大多数的这种材料更换情况。在生产制造开始后出现了可供使用的新材料是更换材料的最常见的原因。当然，这些新材料可能降低成本、改进产品性能。但是，必须对新材料进行认真的评价，以确保其所有性能都被人们所了解。应当时刻牢记，新材料的性能和可靠性很少能像现有材料那样为人们所了解。大部分的产品失效和产品责任事故案件是由于在选用新材料作为替代材料之前，没有真正了解它们的长期使用性能而引起的。

产品的责任诉讼迫使设计人员和公司必须要采用最好的程序进行材料选择工作。在材料选择过程中，五个最常见的问题是：(a)不了解或者未能利用关于材料应用方面的最新和最好的信息资料；(b)未能预见和考虑产品可能的合理用途

(如有可能,设计人员还应进一步预测和考虑由于产品使用方法不当造成的后果。在近年来的许多产品责任诉讼案件中,由于错误地使用产品而受到伤害的原告控告生产厂家,并且赢得判决);(c)所使用的材料的数据不全或者有些数据不确定,尤其是当其长期性能数据是如此的时候;(d)质量控制方法不适当和未经验证;(e)由一些完全不称职的人员选择材料。

通过对上述五个问题的分析,可以得出这些问题是没有充分理由存在的结论。对这些问题的分析和研究可以给避免这些问题的出现指明方向。尽管采用最好的材料选择办法也不能避免发生产品责任诉讼,设计人员和工业界按照适当的程序进行材料选择,可以大大减少诉讼的数量。

从上面的讨论可以看出,选择材料的人们应该对材料的性质、特点和加工方法有一个全面而基本的了解。

第二十三课 车床

车床主要是为了进行车外圆、车端面和镗孔等项工作而设计的机床。车削很少在其他种类的机床上进行,而且任何一种其他机床都不能像车床那样方便地进行车削加工。由于车床还可以用来钻孔和铰孔,车床的多功能性可以使工件在一次装夹中进行几种加工。因此,在生产中使用的各种车床比任何其他种类的机床都多。

车床的基本部件有:床身、主轴箱、尾座、溜板、丝杠和光杠,如图23.1所示。

床身是车床的基础件。它通常是由经过充分正火或时效处理的灰铸铁或者球墨铸铁制成。它是一个坚固的刚性机架,所有其他基本部件都安装在床身上。通常在床身上有内外两组平行的纵向导轨。有些制造厂对全部四条导轨都采用导轨尖顶朝上的三角形导轨(即山形导轨),而有的制造厂则在一组中或者两组中都采用一个三角形导轨和一个矩形导轨。导轨要经过精密加工,以保证其直线度精度。为了抵抗磨损和擦伤,大多数现代机床的导轨是经过表面淬硬的,但是在操作时还应该小心,以避免损伤导轨。导轨上的任何误差,常常意味着整个机床的精度遭到破坏。

主轴箱安装在内导轨的固定位置上,一般在床身的左端。它提供动力,并可使工件在不同的速度下回转。它基本上由一个安装在精密轴承中的空心主轴和一系列变速齿轮——类似于卡车变速箱——所组成。通过变速齿轮,主轴可以在许多种转速下旋转。大多数车床有 8~18 种转速,一般按等比数列排列。一种正在不断增长的趋势是通过电气的或者机械的装置进行无级变速。

由于车床的精度在很大程度上取决于主轴,因此,主轴的结构尺寸较大,通常安装在预紧后的重型圆锥滚子轴承或球轴承中。主轴中有一个贯穿全长的通孔,长棒料可以通过该孔送料。主轴孔的大小是车床的一个重要尺寸,因为当工件必须通过主轴孔供料时,它确定了能够加工的棒料毛坯的最大尺寸。

卡盘安装在主轴上,用来夹紧和带动工件转动。三爪"自定心"卡盘(图23.2a)在夹紧和松开工件时,所有的卡爪都同时移动,它适用于横截面为圆形或六角形的工件。四爪"单动"卡盘(图 23.2b)通过每个卡爪的单独移动将工件夹紧。这种卡盘对工件的夹紧力比较大而且能够对非圆形(正方形,长方形)的工件进行精确定心。

尾座主要由三部分组成。底板与床身的内导轨配合,并可以在导轨上做纵向移动。底板上有一个可以使整个尾座夹紧在任意位置上的装置。尾座体安装在底板上,并能在底板上横向移动,使尾座能与主轴箱中的主轴对正。尾座的第三个组成部分是尾座套筒。它是一个直径通常大约在 51 ~ 76 mm(2 ~ 3 英寸)之间的钢制空心圆柱体。通过手轮和螺杆,尾座套筒可以在尾座体中纵向移入和移出几英寸。

三爪和四爪卡盘通常只适用于短工件的装夹。长工件采用双顶尖装夹。如图 23.3 所示,在工件的每个端面上都加工出一个中心孔。安装在主轴和尾座中的顶尖通过这些中心孔确定了工件轴线的位置。然而,这些顶尖并不能将主轴的运动传递给工件。为此,通常采用拨盘和卡箍来带动工件旋转。在图 23.3 中,安装在主轴内的顶尖为普通顶尖,而安装在尾座内的顶尖为活顶尖。顶尖的柄部通常被加工成莫氏锥度,用来与主轴或尾座中的锥孔相配合。

车床的加工范围可以用两个尺寸参数表示。第一个尺寸参数称为最大回转直径。它大约是两顶尖连线与导轨上最近点之间距离的两倍。第二个尺寸参数是两顶尖之间的最大距离。最大回转直径表示在这台车床上能够车削的工件的最大直径,而两顶尖之间的最大距离则表示在两个顶尖之间能够安装的工件的最大长度。

普通车床是生产中最经常使用的车床种类。它们是具有前面所叙述的所有那些部件的重载机床,并且除了小刀架之外,全部刀具的运动都有机动进给。它们的最大回转直径通常为 305 ~ 610 mm(12 ~ 24 英寸);两顶尖之间距离通常为 610 ~ 1219 mm(24 ~ 48 英寸)。但是,最大回转直径达到 1270 mm(50 英寸)和两顶尖之间距离达到 3658 mm(12 英尺)的车床也并不少见。这些车床大部分都有切屑盘和一个安装在内部的冷却液循环系统。小型的普通车床——最大回转直径一般不超过330 mm(13 英寸)——其中一些也可以被设计成台式车床,即床身可安装在工作台上。

虽然普通车床有很多用途,是很有用的机床,但是更换和调整刀具以及测量工件花费很多时间,所以它们不适合在大批量生产中应用。它们的实际切削加工时间通常少于其总加工时间的30%。

第二十五课　机械加工

机械加工可以被定义为：以切屑的形式从工件上去除不需要的材料（加工余量），以获得成品零件所应该有的尺寸，形状和表面质量的过程。机械加工包括车削，铣削，磨削，钻削等。

车削

普通车床　普通车床作为最早的金属切削机床中的一种，具有许多有用的和为人们所需要的特性。现在，这些机床主要用在规模较小的工厂中，进行小批量的生产，而不是进行大批量的生产。

普通车床的加工偏差主要取决于操作者的技术熟练程度。设计工程师应该认真地确定由熟练工人在普通车床上加工的试验零件的公差。在把试验零件重新设计为生产零件时，应该选用经济的公差。

在现代的生产车间中，普通车床已经被种类繁多的自动车床所取代，诸如转塔车床、自动螺纹车床和自动仿形车床。

转塔车床　对生产加工设备来说，目前比过去更着重评价其是否具有精确的和快速的重复加工能力。应用这个标准来评价具体的加工方法，转塔车床可以获得较高的质量评定。

在为小批量的零件（100～200件）设计加工方法时，采用转塔车床是最经济的。为了在转塔车床上获得尽可能小的公差值，设计人员应该尽量将加工工序的数目减至最少。

自动螺纹车床　自动螺纹车床通常被分为二种类型：单轴自动和多轴自动车床。自动螺纹车床最初是被用来对螺钉和类似的带有螺纹的零件进行自动化和快速加工的。但是，这种车床的用途早就超过了这个狭窄的范围。现在，它在许多种类的精密零件的大批量生产中起着重要的作用。

工件的数量对采用自动螺纹车床所加工的零件的经济性有较大的影响。如果工件的数量少于1000件，在转塔车床上进行加工比在自动螺纹车床上加工要经济得多。如果计算出最小经济批量，并且针对工件批量正确地选择机床，就会降低零件的加工成本。

自动仿形车床　因为零件的表面粗糙度在很大程度上取决于工件材料、刀具、进给量和切削速度，采用自动仿形车床加工所得到的最小公差不一定是最经济的公差。

在某些情况下，在连续生产过程中，只进行一次切削加工时的公差可以达到±0.05 mm。对于某些零件，槽宽的公差可以达到±0.125 mm。在希望获得最大产量的大批量生产中,进行直径和长度的车削时的最小公差值为 ±0.125 mm是经济的。

铣削

除了车削和钻削,铣削无疑是应用最广泛的金属切削方法。铣削非常适合于而且也易于应用在任何数量的零件的经济生产中。在产品制造过程中,许许多多种类的铣削加工是值得设计人员认真考虑和选择的。

与其他种类的加工一样,对于进行铣削加工的零件,其公差应该被设计成铣削生产所能达到的经济公差。如果零件的公差设计得比需要的要小,就需要增加额外的工序,以保证获得这些公差——这将增加零件的成本。

磨削

磨削是一种应用最为广泛的零件精加工方法,用来获得非常小的公差和非常低的表面粗糙度。用于精加工的磨削方法主要有五种:平面磨削(图 25.2),外圆磨削(图 25.3),内圆磨削(图 25.4),无心磨削和砂带磨削。

目前,几乎存在着适合于各种磨削工序的磨床。零件的设计特征在很大程度上决定了需要采用的磨床的种类。当加工成本太高时,就值得对零件进行重新设计,使其能够通过采用既便宜又具有高生产率的磨削方法加工出来。例如,在有可能的情况下,可以通过对零件的适当设计,尽量用无心磨削加工,以获取较好的经济效益。

磨床有以下几种类型:平面磨床(图 25.5)、外圆磨床(图 25.6)、无心磨床、内圆磨床和工具磨床。

平面磨床用来对由各种平面组成的工件,或者带有平面的工件进行精加工。可以采用砂轮的圆周面或者砂轮的端面进行磨削。这类机床上装有往复式工作台或者回转式工作台。

外圆磨床和无心磨床是用来磨削圆柱形工件或者圆锥形工件的。因此,轴和其他类似的零件可以采用外圆磨床或者无心磨床进行加工。

内圆磨床用来磨削精密的孔、汽缸孔以及各种类似的需要进行精加工的孔。

螺纹磨床用来磨削螺纹量规上的精密螺纹和用来磨削螺纹的中径与轴的同心度公差很小的精密零件上的螺纹。

第二十七课　金属切削加工

金属切削加工在制造业中得到了广泛的应用。其特点是工件在加工前具有足够大的尺寸,可以将工件最终的几何形状尺寸包容在里面。不需要的材料以切屑、颗粒等形式被去除掉。去除切屑是获得所要求的工件几何形状,尺寸公差和表面粗糙度的必要手段。切屑的数量多少不一,可能占加工前工件体积的百分之几到 70%～80% 不等。

由于在金属切削加工中,材料的利用率相当低,加之预测到材料和能源的短缺以及成本的增加,最近十年来,金属成形加工的应用越来越多。然而,由于金属成形加工的模具成本和设备成本仍然很高,因此尽管金属切削加工的材料消耗较高,在许多情况下,它们仍然是最经济的。由此可以预料,在最近几年内,金属切削加工在制造业中仍将占有重要的位置。而且,金属切削加工的自动生产系统的发展要比金属成形加工的自动生产系统的发展要快得多。

金属切削加工以切屑的形式从工件表面去除材料。这需要刀具材料的硬度比工件材料的硬度高。因此,刀具的几何形状和刀具与工件的运动方式决定了工件的最终形状。这个基本过程是机械过程:实际上是一个剪切与断裂相结合的过程。

如前所述,在金属切削加工中,多余的材料由刚性刀具切除,以获取需要的几何形状、公差和表面粗糙度。属于此类加工方法的例子有车削、钻削、铰孔、铣削、牛头刨削、龙门刨削、拉削、磨削、珩磨和研磨。

大多数切削加工(或称机械加工)过程是以两维表面成形法为基础的。也就是说,刀具与工件材料之间需要两种相对运动。一种定义为主运动(切削速度主要是由它确定的),另一种定义为进给运动(向切削区提供新的加工材料)。

车削时,工件的回转运动是主运动;龙门刨床刨削时,工作台的直线运动是主运动。车削时,刀具连续的直线运动是进给运动;而在龙门刨床刨削中,刀具间歇的直线运动是进给运动。车削加工的基本参数如图 27.1 所示。

切削速度 切削速度 v 是主运动中刀具(在切削刃的指定点)相对工件的瞬时速度。

车削、钻削和铣削等加工方法的切削速度可以用下式表示:

$$v = \pi d n \quad \text{m/min} \tag{27.1}$$

式中 v 为切削速度,其单位为 m/min; d 是工件上将要切削部分的直径,其单位为 m; n 是工件或主轴的转速,单位为 r/min。根据具体运动方式不同,v、d 和 n 可能与加工材料或工具有关。在磨削时,切削速度通常以 m/s 为单位度量。

进给量 在主运动之外,当刀具或工件作进给运动 f 时,便产生重复的或连续的切屑切除过程,从而形成所要求的加工表面。进给运动可以是间歇的,或者是连续的。进给速度 v_f 定义为在切削刃的某一选定点上,进给运动相对于工件的瞬时速度。

对于车削和钻削,进给量 f 以工件或刀具每转的相对移动量(mm/r)来表示;对于龙门刨削和牛头刨削,进给量 f 以刀具或工件每次行程的相对移动量(mm/stroke)来表示。对于铣削,以刀具的每齿进给量 f_z(mm/tooth)来表示,f_z 是相邻两齿间工件的移动距离。所以,工作台的进给速度 v_f(mm/min)是刀具齿数 z,刀具每分钟转数 n 与每齿进给量 f_z 的乘积($v_f = nzf_z$)。

包含主运动方向和进给运动方向的平面被定义为工作平面,因为该平面包含决定切削作用的两种基本运动。

切削深度(背吃刀量) 如图 27.1 所示,车削时的切削深度 a(有时也被称为吃刀深度)是刀具切削刃切进或深入工件表面内的距离。切削深度决定工件的最终尺寸。在车削加工中采用轴向进给时,切削深度可以通过直接测量工件半径的减少量来确定;在车削加工中采用径向进给时,切削深度等于工件长度的减少量。

切屑厚度 未变形状态时的切屑厚度 h_1,就是在垂直于切削方向的平面内垂直于切削刃测量得到的切屑厚度(见图 27.1)。切削后的切屑厚度(即切屑实际厚度 h_2)大于未变形时的切屑厚度,也就是说切削比或者切屑厚度比 $r = h_1/h_2$ 总是小于 1。

切屑宽度 未变形状态的切屑宽度 b,是在与切削方向垂直的平面内沿切削刃测得的切屑宽度。

切削面积 对于单刃刀具切削加工,切削面积 A 是未变形的切屑厚度 h_1 和切屑宽度 b 的乘积(即 $A = h_1 b$)。切削面积也可以用进给量 f 和切削深度 a 表示如下:

$$h_1 = f\sin\kappa \quad \text{及} \quad b = a/\sin\kappa \quad (27.2)$$

式中 κ 为主偏角。

因此,可以由下式求出切削面积

$$A = fa \quad (27.3)$$

第二十九课　齿轮制造方法

刨齿 齿轮齿间的空间形状是复杂的,而且随着齿轮的齿数和模数的不同而变化,因此大多数的齿轮制造方法采用展成法加工齿面而不是用成形法加工。

刨齿采用往复运动的齿条刀,当齿条刀实际上绕齿轮坯料滚切并沿其螺旋线方向运动时,齿形就会被逐渐展成。与普通的刨削加工相同,在回程中齿条刀与齿轮脱离接触。这种加工方法的最大好处是,刀具为具有直线齿形或者接近于直线齿形的齿条,其齿面易于进行精确磨削加工。由于加工速度缓慢,这种方法几乎不在大批量的生产中应用。对于单件或者少量的齿轮加工而言,缓慢的行程速度带来的影响不大,而且较低的刀具成本对于那些特殊规格和需要进行齿廓修形的齿轮来说则是一个有利条件。

插齿 插齿加工在本质上与刨齿加工类似,只是采用了圆形刀具来取代齿条刀,如图 29.1 所示。其结果是减少了往复运动惯性,在加工过程中可以采用比刨削高得多的行程速度。现代插齿机在加工汽车齿轮时可以达到每分钟 2000 次切削行程。

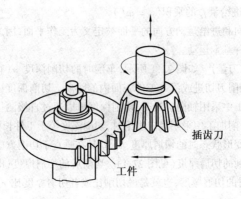

图 29.1 插齿加工

插齿加工时,在刀具的每一次行程中,通常刀具和工件的切向移动距离为 0.5 mm。在回程中,刀具必须退让 1 mm 以留出间隙。否则,在退刀时,刀具会擦伤已加工表面,并且加快刀具的磨损。

插齿加工的优点是生产率较高和可以将齿插到接近轴肩处。令人遗憾的是,加工斜齿轮时,需要有一个能够产生绕插齿运动行程本身旋转的螺旋导轨,如图 29.2 所示。这种螺旋导轨不易制造,或者说其制造成本较高。由于对每一种不同螺旋角的齿轮,应该制造不同参数的插齿刀和螺旋导轨,因此这种方法仅适用于斜齿轮的大批量加工。插齿加工的一大优点是能够加工诸如行星齿轮传动所需要的内齿轮。

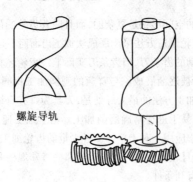

图 29.2 斜齿轮插齿加工

滚齿 滚齿是最常见的金属切削方法。它采用齿条展成原理,但是通过安装在旋转刀具上的许多"齿条"来避免缓慢的往复运动。这些"齿条"轴向排布,形成了开槽的蜗杆(见图 29.3),滚齿加工过程如图 29.4 所示。

图 29.3 齿轮滚刀

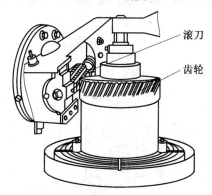

图 29.4 滚齿加工

由于滚刀和工件均不作往复运动,因此滚齿加工时的金属切削率很高。对于普通滚刀可以采用 40 m/min 的切削速度,对于硬质合金滚刀可以采用高达 150 m/min 的切削速度。一般采用直径为 100 mm 的滚刀,其转速为 100 r/min,一个 20 个齿的齿轮将以 5 r/min 的速度旋转。工件的每一转将有 0.75 mm 的进给量,因此滚刀每分钟对工件的进给大约为 4 mm。在汽车制造中,采用多头粗加工滚刀,达到每转 3 mm 的大进给量。这样当刀具的转速为 100 r/min 时,采用双头滚刀加工齿数为 20 的齿轮,其进给量可达到 30 mm/min。

拉齿 通常不采用拉削的方式加工斜齿轮,但是在内直齿轮加工中拉削是适用的。在这种情况下,拉削主要被用来加工其他任何一种方法都不容易加工的内花键。同所有的拉削加工一样,因为设备的费用很高,齿轮拉削方法只有在大批量生产时才是经济的。

通过拉齿可以获得高精度和低表面粗糙度。但是,如同所有的切削加工一样,拉削也只能用来加工"软"材料。随后应该对材料进行表面硬化或者热处理,而这会产生变形。

剃齿 采用剃齿方式可以对处于"软"状态的齿轮进行精加工。其目的是通过采用具有提高齿形精度能力的"刀具"与经过粗加工的齿轮进行啮合来降低齿面粗糙度和改善齿廓形状。

剃齿刀形状如齿轮,在刀齿的根部有一附加的凹槽(用于排除细切屑和冷却液),在齿面上开有许多小槽以形成切削刃。剃齿刀与经过粗加工的齿轮成交错轴啮合传动,如图29.5所示。这样在理论上沿着轮齿有一个具有相对速度的点接触,从而产生刮削作用。剃齿刀的刀齿具有相当的弯曲柔性,因此只有当它们在两个轮齿之间并且与这两个轮齿都接触时,才能有效地进行切削工作。加工周期可能少于半分钟,机床的价格也不昂贵,但是刀具比较精密,难于制造。

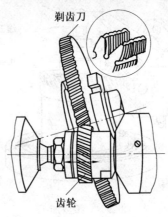

图 29.5 剃齿加工

磨齿 磨齿是非常重要的,因为它是加工淬硬齿轮的主要方法。当对于热处理变形的预先校正达不到齿轮所要求的高精度时,就必须采用磨削加工。

最简单的磨齿方法是成形磨削法。采用由精密切削成形的样板控制的单点金刚石可以将砂轮修整成精确的形状。成形砂轮沿着齿轮轴向往返进给。在一个轮齿的形状加工完成以后,通常要除去 100 μm 的金属。然后,齿轮将会被分度到下一个轮齿空间。这种方法的加工速度相当慢,但是在整个加工过程中都可以获得较高的精度。对于不同的模数、齿数、螺旋角或者齿廓修形量,需要采用不同的砂轮修整样板,因此就需要有较长的安装调整时间。

最快的磨齿方法采用与滚齿相同的原理,但是使用截面为齿条的砂轮来取代开槽的蜗杆形滚刀。砂轮只能被切削成单头蜗杆形状,由于齿轮的转速通常高达 100 r/min,因此设计具有所需要的精度和刚度的驱动系统是困难的。尽管在磨削过程中砂轮和工件有产生不同的变形量的可能,可能需要用砂轮的形状补偿机床变形的影响,磨削加工的精度还是比较高的。将砂轮展成为一个蜗杆形状是一个缓慢的过程,这是因为修整砂轮的金刚石不仅要使砂轮形成齿条外形,而且需要在砂轮旋转时进行轴向移动。一旦砂轮整形完毕后,就能快速地磨削加工齿轮,直至需要重新修整时为止。这是一个最常用的,高效率地加工小齿轮的方法。

第三十一课 尺寸公差与表面粗糙度

现代技术对零件尺寸精度的要求越来越严格。而且,目前许多零件是由散布在各地的不同厂家生产的,因此必须对这些零件的尺寸规格和生产做出严格的规定,以保证它们具有互换性。

给零件标注尺寸使其在一个规定的区间内变动,以保证它们具有互换性的技术称为公差技术。允许每个尺寸在规定范围内具有的一定的变动量,称为公差。例如,一个零件的尺寸可以被表示成 20 ± 0.05,其公差(尺寸变动量)为 0.10 mm。

由于将零件精确地加工到某一指定的尺寸(除了碰巧外)是不可能,而且这样做是既不必要也不经济的,因而相对于所需要的理论尺寸,公差是一个既不希望存在,但又是允许存在的偏差。

在不影响零件性能的情况下,应当给尺寸尽可能大的公差,以利于把生产成本降至最低。制造成本会随着公差的降低而升高。

有三种表示尺寸公差的方式:单向,双向和极限方式。当采用正负公差时,就将公差加到被称为基本尺寸的理论尺寸上去。当只允许尺寸向基本尺寸的单一方向(或者变大,或者变小)的变动时,就是单向公差。当尺寸在基本尺寸的两个方向(变大或者变小)都可以变动时,公差就是双向。公差也可以用极限形式给出,表示零件外形的最大和最小尺寸。

一些与公差有关的术语和定义如下述:

公差:是为某一个尺寸所规定的上限与下限之间的差值。

基本尺寸:理论尺寸,是计算极限尺寸和偏差的起始尺寸。

偏差:孔的尺寸或者轴的尺寸减去基本尺寸所得的差值。

上偏差:零件最大极限尺寸减去其基本尺寸所得的差值。

下偏差:零件最小极限尺寸减去其基本尺寸所得的差值。

实际尺寸:加工后零件的实测尺寸。

配合:两个装配在一起的零件之间的松紧程度。可以把配合分为三类:间隙配合,过盈配合,过渡配合。

设计工程师并不需要知道如何操纵一台专门的机床去获得所需要的表面粗糙度,但是他应该清楚这些加工方法的某些情况。各种工序所能得到的表面粗糙度数值的大小,以及采用每一种工序来获得更为光滑表面的经济性方面的知识都会帮助设计工程师决定采用何种表面粗糙度。

由于其简单,算术平均粗糙度 Ra 在全世界范围内得到了广泛的应用。下面各段中介绍了表面粗糙度 Ra 的应用实例。

0.2μm 这种表面粗糙度适用于液压支柱的表面,液压缸,安装有 O 型密封圈的活塞和活塞杆,滑动轴承中的轴颈,凸轮表面,承受正常载荷的滚动轴承中的滚柱等。

0.4μm 这种表面粗糙度适用于高速转动的轴承,承受重载的轴承,普通商业级轴承的滚柱,液压零部件,密封环槽底,在滑动轴承中工作的轴颈和承受很高拉应力的构件。

0.8μm 这种表面粗糙度通常见于有应力集中和承受振动载荷的零件,以及拉削的孔,齿轮齿面和其他精密加工的零件。

1.6μm 这种表面粗糙度适合于普通轴承,一些尺寸公差较小的普通机械零件和不承受剧烈交变应力的高强度零件。

3.2μm 这种表面粗糙度不能用于滑动表面,但可以用于承受较低载荷和不经常承受载荷的粗糙支承面,或者适用于承受中等应力的机械零件。

6.3μm 这种表面粗糙度的外观并不太难看,可以用于非关键零件的表面和支座的安装面等等。

由于大多数制造行业均具有高度的竞争性,因此寻找降低成本的途径这个问题一直是人们所关心的。降低成本的最好起点是在产品的设计阶段。设计工程师在进行设计时应该记住所有可供选择的方案。如果没有对可能的生产成本进行认真的分析,通常不可能选定最好的方案。在对产品的功能、互换性、质量和经济性进行设计时,需要对其公差、表面粗糙度、加工方法、材料以及设备等问题进行认真的研究。

对于互换性原理的全面、深入地研究是充分理解和正确认识低成本生产技术的基本因素。互换性原理是保证任何数量的零件能够顺利生产的关键。应该认真地综合考虑所有零件的细节,以保证不仅可以采用低成本的加工工艺,而且可以实现快速和容易的装配和维修。

第三十三课　产品图样

制造企业的成立是为了生产一种或多种产品。这些产品是采用被称为产品图样的文件来进行完全定义的。产品图样上的尺寸和技术要求将保证零件具有互换性和能够可靠地实现其设计性能。产品图样通常包括零件图和装配图。

零件图是一个零件的标注有尺寸的多个视图,它描述了零件的形状、尺寸、材料和表面粗糙度,这张图样本身含有制造这个零件所需要的全部信息。大部分零件需要采用三个视图对其形状做完整的描述。

装配图表明设计中的所有零件是如何被装配到一起的,以及整个装置的功能;因此,在其中完整的形状描述并不重要。应该采用必要和尽可能少的视图对装配体中各个零件之间的相互关系进行描述。

所有的零件都必须在某种程度上与其它零件相互作用,以产生设计方案中期望得到的功能。在绘制零件图之前,设计人员应该对装配图进行透彻地分析,以保证零件之间有适当的配合,所标注的公差准确无误,接触表面经过适当的机械加工,零件之间可以产生正确的运动。

英寸是英制的基本单位,实际上在美国所有的制造图样中都是用英寸来标注尺寸。

毫米是公制的基本单位。标注尺寸时可以略去数字后面公制单位的缩写mm,因为标题栏附近的 SI 符号表明所有的单位都是公制的。

在美国,有些图样同时用英寸和毫米两种单位标注尺寸,通常将用毫米标注的尺寸放在圆括号或方括号中。也可以先用毫米作为单位标注尺寸,然后将尺寸单位换算成英寸,并将其放在方括号中。两种单位之间的换算会产生必须舍入的小数位误差。每张图样所用的主要单位制,必须在标题栏中加以说明。

标题栏 在实践中,标题栏中通常包括图样名称或零件名称、制图员、日期、比例、单位名称和图样代号。另外,还可以包含审核人员和材料等信息。在第一次制图之后,任何为了改进设计方案而作的变更和改进都应该在更改区中标明。标题栏一般位于图纸的右下角。

比例是图样中的图形与实物相应要素的线性尺寸之比。不论采用什么比例,在图样上的尺寸都是指物体的真实尺寸,而不是视图的尺寸。在一套产品图样中可以采用各种不同的比例。

根据项目复杂程度的不同,一套产品图样中图样的数量少则一张,多至一百张以上。因此在每张图样上都必须写上这套图样的总张数和该张图样所在的张次(例如,共 6 张第 2 张,共 6 张第 3 张,等等)。

零件名称和序号 给每个零件一个名称和序号,通常采用的字母和数字的高度为 1/8 英寸(3 毫米)。零件的序号应放在该零件视图的附近,以便清楚地表明它们之间的关系。

明细栏 在明细栏中,零件的序号和名称必须与产品图样中的同一个零件一致。此外,还要给出所需要的同样零件的数量和制造零件所用的材料。

一直到 1900 年,世界各国的图样通常都采用第一角投影的画法绘制。在第一角投影中,俯视图放在主视图的下面,左视图放在主视图的右面,依此类推。现在,美国、加拿大和英国通常采用第三角投影,而世界上许多国家仍然采用第一角投影。在第三角投影中,俯视图放在主视图的上面,右视图放在主视图的右面,左视图放在主视图的左面。

实际上,第一角投影与第三角投影的唯一区别就是视图的位置。当使用者阅读第一角投影图样而将其当作第三角投影图样时会产生混乱和制造误差,反之亦然。为了避免误解,可以采用投影识别符号(见图33.1)来区分第一角投影与第三角投影。如果某些图样的投影方式可能会引起误解,这些符号就会出现在标题栏中或者靠近标题栏的位置。

产品图样是法律合同,它记录了在设计工程师指导下制订的设计细节和技术要求。因此,图样必须尽可能地清楚、准确和详尽。对于一个项目,在生产时进行修改或改进所付出的代价要远比在设计初始阶段进行这项工作要花更多的费用。为了在经济上有竞争优势,图样必须尽可能地没有差错。

图样的审核人员应该具有特殊的素质,他们能发现图中的错误,并提出修改和改进意见,以便制造出物美价廉的产品。审核人员通过装配图和零件图来检查设计方案是否正确无误。此外,他们还负责检查图样的完整性、质量和清晰度。

除了对每处修改都要做好记录外,制图员应该记录下整个项目期间所作的一切更改。随着项目的进行,制图人员应该将所有的更改,日期和有关人员的姓名记录下来。这样的记录使得任何人在将来对项目进行复核时,会很容易和很清楚地了解为了获得最后设计方案所经历的过程。

第三十五课　特种加工工艺

人类通过使用工具和智能来制造使其生活变得更容易和更舒适的物品这种方法,把他们自己与其他种类的生命区别开来。许多世纪以来,工具和为工具提供动力能源的种类都在不断地发展,以满足人类日益完善和越来越复杂的想法。

在最早的时期,工具主要是由石器构成。考虑到所制造的物品相对简单的形状和被加工的材料,石头作为工具是适用的。当铁制工具被发明出来以后,耐用的金属和更精致的物品能够被制造出来。在20世纪中,已经出现了一些由有史以来最耐用,同时也是最难加工的材料制造的产品。为了迎接这些材料给制造业带来的挑战,工具材料已经发展到包括合金钢、硬质合金、金刚石和陶瓷。

给我们的工具提供动力的方法也发生了类似的进步。最初,是由人或动物的肌肉为工具提供动力。随后,水力、风力、蒸汽和电力得到了利用,人类能够通过采用新型机器、更高的精度和更快的加工速度来进一步提高制造能力。

每当采用新的工具、新的材料和新的能源时,制造效率和制造能力都会得到很大的提高。然而,当旧的问题解决了之后,就会有新的问题和挑战出现。例如,现今制造业面对着下面一些问题:你如何去钻一个直径为2 mm,长度为670 mm的孔,而不产生锥度和偏斜?用什么办法能够有效地去除形状复杂的铸件内部的通道中的毛刺,而且保证去除率达到100%?是否有一种焊接工艺,它能够避免目前在我的产品中出现的热损伤?

从20世纪40年代以来,制造业中发生的大变革一次又一次地促使制造厂家去满足日益复杂的设计方案和很高耐用度,但是在许多情况下几乎接近无法加工的材料所带来的各种要求。这种制造业的大变革不论是现在还是过去都是集中在采用新型工具和新型能源上。这样做的结果是产生了用来去除材料、成型、连接的新型制造工艺。这些工艺目前被称为特种加工工艺。

在目前所采用的常规制造工艺中,材料的去除是依赖于电动机和硬的刀具材料进行的,诸如锯断、钻孔和拉削。常规的成型加工是利用电动机、液压和重力所提供的能量进行的。同样,材料连接的常规做法是采用诸如燃烧的气体和电弧等热能进行的。

与之相比,特种加工工艺采用按照以前的标准来说不是常规的能源。现在材料的去除可以利用电化学反应、电火花(图35.1)、高温等离子、高速液体和磨料射流。过去非常难进行成型加工的材料,现在可以利用大功率的电火花所产生的磁场、爆炸和冲击波进行成型加工。采用高频声波和电子束可以使材料的连接能力有很大的提高。

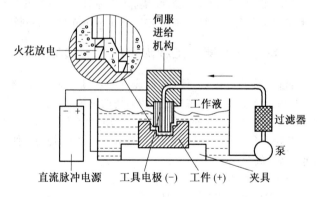

图35.1 电火花加工过程示意图

在过去的50年间,人们发明了20多种特种加工工艺,并且将其成功地应用于生产之中。这么多种特种加工工艺存在的原因与许多种常规加工工艺存在的原因是一样的。每一种工艺都有它自己的特点和局限性。因而,不存在一种对任何制造环境来说都是最好的工艺方法。

例如,有时特种加工工艺或者通过减少生产某一产品所需要的加工工序的数量,或者通过采用比以前使用的方法更快的工序来提高生产率。

在另外的场合中,采用特种加工工艺可以通过增加重复精度,减少易损坏工件在加工过程中的损伤,或者减少对工件性能的有害影响来减少采用原来加工工艺所产生的废品数量。

由于前面所提到的这些特点,特种加工工艺从其诞生时起就开始了稳定的发展。由于下列原因,可以肯定这些工艺将来会有更快的增长速度:

(1)目前,同常规工艺相比,除了材料的体积去除率外,特种加工工艺几乎具有不受限制的能力。在过去几年中,某些特种加工工艺在提高材料去除率方面有了很大的进展,而且有理由相信这种趋势在将来也会继续下去。

(2)大约半数的特种加工工艺目前采用计算机控制加工参数。使用计算机可以使人们所不熟悉的加工过程变得简单,因而加大了人们对这种技术的接受程度。此外,计算机控制可以保证可靠性和重复性,这也加大了人们对这种技术的接受程度和其应用范围。

(3)大多数特种加工工艺可以通过视觉系统,激光测量仪表和其他加工过程中的检测技术来实行适应控制。例如,如果加工过程中的检测结果表明,产品中正在加工的孔的尺寸在变小,可以在不更换硬的加工工具(如钻头)的情况下,修正孔的尺寸。

(4)随着制造工程师,产品设计人员和冶金工程师们对特种加工工艺所具有的独特能力和优越性的了解的增加,特种加工工艺的应用范围将会不断增加。

第三十六课 工程陶瓷的机械加工

工程陶瓷材料具有许多引人注目的特性,诸如:高硬度,高耐热性,化学惰性,低的导热性和导电率等。然而,这些特性使得陶瓷的机械加工非常困难,不管是采用磨削或者非磨削的方法都是如此。当采用非磨削加工方法时,必须提高加工生产率才能在经济上合算。其他要求包括:降低表面粗糙度,控制工件的几何形状特征,降低设备的成本。

磨削涉及到工件的材料性能、砂轮规格、机床的选择与砂轮准备等许多变量之间复杂的相互影响。这种相互影响叫做"磨削性",它可以用材料去除速度、所需要的动力或者磨削力、表面粗糙度、公差和表面完整性等项指标的数量大小来表示。

在磨削过程中,工件表面上会产生一些影响表面完整性的缺陷,在每个工件上所产生的缺陷的大小和多少都与在其他工件上的不相同。这就造成了一批工件的断裂强度值的大小不同。机械加工中的切屑去除过程,也会在表面上和表面层下面产生残余应力。表面层中的缺陷的长度可能在 $10 \sim 100\ \mu m$ 之间。这些损伤通常应该采用细砂轮精磨的方式除去。

对于陶瓷的常规磨削加工,磨床与砂轮的选择也会影响磨削性。一般来说,只能采用金刚石砂轮加工陶瓷。加工烧结后的陶瓷,用树脂结合剂的金刚石砂轮要比用金属或陶瓷结合剂的砂轮的效果要好。树脂结合剂砂轮产生的磨削力较小,磨损较快,比较容易露出新的磨粒,它可以将陶瓷工件的表面缺陷尺寸减到最小。

因为陶瓷的性能对最终的磨削结果起重要作用,必须深入了解它们与机械加工变量之间的相互作用,以便确定成本。对于导热性低的材料,冷却液的应用是非常关键的,它可以防止产生热致裂纹。孔隙率、粒度与显微组织都会影响表面粗糙度和表面质量;孔隙率高会产生较差的表面粗糙度。一般来说,加工陶瓷需要较高的磨削力和功率,这会缩短砂轮的使用寿命。

有几种磨削方法不需要使用昂贵的金刚石砂轮,因而不存在由金刚石砂轮带来的问题。水射流加工是采用高速液体射流来去除材料。射流可以是单独使用水流或者是在水中夹带磨料粒子。射流可以有脉冲或连续两种方式,对大多数陶瓷的加工采用连续射流方式。这种加工方式最适用于切削缝隙、沟槽或者大孔的套料加工。水射流加工既无热影响也不需要大的机械力,因此在切削对热冲击最敏感的和最脆弱的陶瓷时,都不会造成工件的损坏。这种加工方法可以避免产生颤振、振动、工件的表面变形或表面层下的损伤。

超声波加工是另外一种磨削加工过程,它具有一些优于常规磨削的特点。超声加工有时被称为冲击磨削,它是使用高频换能器的一种机械加工过程。当电能转换为机械运动时,就会产生小振幅的振动。振动传到工具柄,使其以每秒 20000 次的频率产生通常为 0.02 mm 的直线机械运动。

这种作用以稳定的进给率实现微量切削。在受到控制的压力下使用稳定流率的磨料悬浮液。最常用的磨料是 20%~50% 浓度的碳化硼或碳化硅。悬浮液是循环使用的,可不断提供新磨料和除去磨碎的颗粒。磨料的粒度决定了工件的表面粗糙度、相对于工具的空隙尺寸与切削速率。可以采用真空来提高磨料的流动率。工具的形状会反映到被加工的工件上。可以通过增大工具柄的振幅来增大行程,从而提高切削生产率。

超声波加工几乎可以用来加工任何一种硬脆材料,但是加工 40 HRC 以上的材料时更为有效。这些材料包括硅、玻璃、石英、光纤材料、结构陶瓷(如碳化硅)和电子基片(如氧化铝)。

非磨削加工,诸如激光束加工和电子束加工,不受陶瓷硬度的限制,因此可以用来代替常规磨削加工。在激光束加工中,激光器将能量密度高的光束聚焦在工件上,使工件材料蒸发。由于激光束是由机械定位的,因此没有电子束加工快。但与电子束加工不同的是不需要真空室,因而工件的装入较快,并且不受尺寸限制。另一个优点是设备的价格较低。通常采用脉冲方式钻孔，采用连续的方式切削。激光束可以用来加工任何硬质材料，其中包括金刚石。

加工速度是由激光使材料熔化、气化蒸发并去除的速度所决定的。材料的去除是热对流和光束的压力所引起的。如果采用气流吹走熔化、气化的材料，就可以显著地提高材料的去除率。热梯度能够引发裂缝。因此，厚陶瓷应该在尚未烧结的状态下加工。相反，因为加热能产生有益的残余应力，故经过激光加工的陶瓷的强度可以高于经过金刚石加工的陶瓷。根据需要去除的材料量来确定激光器的功能，一台150瓦的激光器就可以用来加工薄陶瓷，而加工厚陶瓷则通常需要采用15000瓦的激光器。

第三十七课 机械振动

机械振动是物体和结构的连续的或者瞬态的振荡运动。在某些情况下，在设计机器时就考虑到使其产生机械振动，并将振动作为机器功能的一部分，属于这类机器的有风钻、往复式发动机等。然而，在大多数情况下，振动是偶然或意外产生的，它们会影响结构或仪器的正常功能。

由于机械振动已经渗透到机械工程的各个方面，因而在某种程度上成为工程和物理领域中共同关心的一门学科。所以，从事各种专业的工程技术人员都应该掌握机械振动的基本知识。

振动在机械系统中的作用 有许多原因使得人们对机械振动的基本理论和实际应用产生了广泛的兴趣。

这样做的原因之一是振动会对机械系统产生不良的影响。一般的机械系统都可以简化为由质量、弹簧、阻尼器这些元件以某种方式相互连接所组成的系统模型。整个建筑物、放在实验室工作台上的仪器、安置在车间地面上的复杂机床、运输车辆以及人的身体都是一般机械系统的例子。由于大多数激振力函数 $f(t)$ 中含有谐波分量，所以引起整个系统产生共振的可能性很大。如果产生共振的系统是无阻尼的，那么质量的位移和弹簧的伸长量将趋于无穷大。这将会使弹簧发生断裂，因此为了保护设备和仪器，必须避免产生无阻尼共振。这也适用于人的身体作为整个系统的一部分时的情况，系统发生共振时会对人体造成伤害。

即使机械系统长期处于振动频率远离其共振频率的振动环境中，也会使系统产生疲劳破坏。因为像弹簧这样的机械零件，在比极限强度低很多的重复应力或周期性应力的作用下，经过很多次的重复作用之后也将发生断裂。实际上如果增加应力周期的次数，那么，能够最终引起断裂的应力值将减小。疲劳的基本理论认为，分子之间的键是从分子结构有缺陷或者薄弱的地方被逐渐拉开的。

振动的另一个不良影响是它们能够损害仪器的正常功能。如果一台放大倍数

超过 10^4 倍的电子显微镜内发生振动,那么得到的会是模糊的图像。切片机内的振动会使切片的厚度不同。同样的原因,许多精密工程和光学仪器与装置都不允许发生过大的振动。振动能够使电器设备中的连接部分松开。

系统中产生不需要的振动还会降低系统的效率。因为产生振动会浪费一部分原本是用来带动系统正常工作的能量。

结构发生振动的另一个不良的副作用是会产生刺耳的噪声。这种噪声会使处于这个环境中工作的人们在精神上感到烦躁,并会影响正常声音的传播。在极端情况下,噪声会使人的听力受到永久性伤害。抑制噪声最有效的措施是减少或者消除产生噪声的振动。

人们在与地震有关的地面振动的测量和研究方面做了很多工作。这种测量是提供预报和保护人们免遭与地面振动有关的火山爆发的伤害的一个重要环节。

另一个使人们感兴趣的振动的定量化研究领域是机械设备,特别是旋转式机械的定期保养或者预防性检修。伴随着这种机械的老化和逐渐磨损,随之而来的有害的振动会逐渐增大。定期进行的振动测量可以提供机械设备使用期间的性能降低指数。据此,可以在对工厂来说是方便的时间里对设备进行检修和更换,以避免发生设备的突发性损坏。

在振动理论所涉及的上述任何领域中,首要的工作是对振动进行测量。

测量设备 最常用的振动测量方法是电测法。这种方法的基本组成元件如图 37.1 所示。其中,振动传感器是关键组件,它能够产生与机械振动的位移、速度或者加速度等的数值成正比的电压或电流。然后,采用各种电子元件对振动电压实施标准电子信号处理。典型的处理步骤有放大、缩小、滤波、微分和积分。随后,用仪器测量被处理过的信号,用示波器进行显示,用曲线记录仪或者磁带记录仪进行记录,或者采用数字计算机作进一步处理和分析。

图 37.1 对振动进行检验和测量的系统示意图

第三十八课 振动的定义和术语

所有的物质——固体、液体和气体——都能够产生振动,例如,在喷气发动机尾管中产生的气体振动会发出令人讨厌的噪声,而且有时还会使金属产生疲劳裂缝。液体中的振动总是纵向的,而且由于液体的可压缩性低,这种振动还会产生很大的力,例如,当输水管道的阀门或水龙头突然关闭时,管道会遭受很大的惯性力的作用(或称为水击)。由于液体流动的状态改变,不平衡的回转件的转动和往复运动零件所产生的激振力,一般可以通过精心设计和制造来使其得到降低。一台常见的机器中有许多运动零件,每个零件都是潜在的振动源或冲击激振源。设计人员需要处理好振动与噪音的允许值与降低激振所需要的费用之间的关系。

机械振动或者是由稳态的谐振力引起的振动(也就是服从正弦或余弦定律的强迫振动),或者是在初始扰动之后,除了被称为重量的重力之外,没有其它外力引起的振动(也就是自然或自由振动的情况)。如果仅用一个频率的正弦或余弦波图形就可以表示位移与时间的关系,谐和振动就被认为是"简单的"。

一个物体或一种材料在振动时,它相对于静平衡位置的位置变化或位移是周期性的。与振动有关的几个相互关联的物理量是加速度,速度和位移。例如,一个不平衡的力在系统中产生的加速度($a=F/m$)会因为系统的抵抗而引起振动作为响应。振动或者振荡运动大致可以分为三类:(1)瞬态的,(2)连续的或稳态的,(3)随机的。

瞬态振动 是逐渐衰减的,而且通常与不规则的扰动有关,例如,撞击或碰撞力,滚动载荷通过桥梁,汽车通过坑洞,也就是在确定的期间内不重复的力。尽管瞬态振动是振动的暂时性成分,它们能够产生大的初始振幅和引起高的应力。但是,在大多数情况下,它们持续的时间很短,因而人们可以将其忽略不计而只考虑稳态振动。

稳态振动 通常和机器的连续运转有关,而且尽管这种振动是周期性的,但不一定是谐和振动或正弦振动。由于需要能量才能产生振动,因此,振动消耗了能量,降低了机器和机构的效率。能量的消耗有多种方式,例如,摩擦和由其产生的热量传递到周围环境,声波和噪音,通过机架与基础的应力波等。因此,稳态振动总是需要连续的能量输入来维持其存在。

随机振动 是一个用来描述非周期性振动的术语。也就是说,这种振动不是周期性变化的,是不定期地进行重复的。

在下面的段落中,对一些与振动有关的术语和定义加以明确,其中一些可能是工科学生都已经清楚了的。

周期,循环,频率和振幅 稳态机械振动是系统在一定时间范围内的重复运动,该时间范围被称为周期。在任何一个周期内所完成的运动,被称为一个循环。每个单位时间内的循环数目被称为频率。系统任何部分离开它的静平衡位置的最大位移就是该部分振动的振幅,总的行程是振幅的两倍。因此,"振幅"并不是"位移"的同义词,而是距离静平衡位置的最大位移。

自由振动和强迫振动 除了重力以外,在没有任何其它作用力时产生的振动称为自由振动。通常一个弹性系统离开它的稳定平衡位置后并被松开时,这个系统就会产生振动,也就是说,自由振动是在弹性系统固有的弹性恢复力的作用产生的。而固有频率则是系统的一个特性。

强迫振动是在外力的激励(或者外部施加的振荡性干扰)下所产生的。这个激励或干扰通常是时间的函数。例如,在不平衡的转动部件中,或者是在有缺陷的齿轮和传动装置中就会产生这种振动。强迫振动的频率就是激励力或者外部施加的力的频率。也就是说,强迫振动的频率是一个同系统固有频率没有关系的任意量。

共振 共振描述了最大振幅的状况。当外力的频率与系统的固有频率相同或相近时就会产生共振。在这种临界条件下,机械系统中会出现具有危险性的大振幅和高应力。但是,电学上,收音机和电视机的接收器则被设计成在共振频率时工作。因此,在所有各种振动或振荡系统中,计算或者估计出系统的固有频率是非常重要的。当转轴或主轴发生共振时,这时的转动速度称为临界转速。在没有阻尼或者其它振幅限制装置的情况下,共振就是在有限激振下产生无穷大响应的条件。因此,预测、修改或消除机械装置产生共振的条件是非常重要的。

阻尼 阻尼是振动系统中能量被消耗的现象,它可以防止过量的响应。可以观察到,自由振动的振幅会随时间而衰减,因而,振动最终将由于某些限制或阻尼的影响而停止。因此,如果要使振动持续下去,一定要有外部的能源对由于阻尼而耗散的能量进行补充。

能量耗散以某种方式与系统的部件或元件之间的相对运动有关,它是由于某种类型的摩擦引起的。例如,在结构中,材料内部的摩擦和由空气或液体阻力等造成的外部摩擦被称为"粘性"阻尼,在这里假定阻力与运动部件之间的相对速度成正比。一种能够提供粘性阻尼的装置被称为"阻尼器"。它是由一个缸体与一个活塞松弛配合所形成的,液体能够从活塞的一端通过环形间隙流到另一端。阻尼器不能存储能量,仅能消耗能量。

第三十九课 残余应力

残余应力是结构或者机器中在没有外部载荷或者内部温差时存在的一种应力。它通常是在制造或者装配过程中所产生的。当这些结构或者机器投入使用时,工作载荷就会与残余应力相叠加。如果残余应力与工作应力相加,则这种残余应力是有害的;如果残余应力与工作应力相减,则这种残余应力是有益的。

这里仅列举几个有害的残余应力的例子。在机器的装配过程中,当两轴不在一条直线上时,采用刚性联轴器强行将它们连接在一起就会产生有害的残余应力。当轴旋转时,轴上所产生的应力就是一个方向不断改变的应力。通常情况下,在无法经济地实现两轴线之间的精确找正对准时,应该使用能够补偿一定找正误差的弹性连轴器。

不均匀加热或者冷却通常会产生残余应力。焊缝是一个最常见的例子。在焊缝凝固后,焊缝金属和邻近区域的温度要比金属主体的温度高很多。金属沿着焊缝长度的自然收缩被邻近的体积较大、温度较低的金属限制了一部分。因此沿着焊缝产生了残余拉应力。

一般规律是"最后冷却的部位处于受拉状态",但是如果显微组织发生某种变化就会出现例外的情况。将这些残余拉应力减至最小或者使之反向的方法有:通过退火消除应力,对强度降低的表面进行锤击或喷丸处理。在退火过程中,要求把低碳钢加热到 600~650℃,某些合金钢要加热到 870℃,然后保温一段时间,随后进行缓慢冷却。对被连接零件进行某种预热可以将焊缝中的拉应力降至最低。

滚压、挤孔和喷丸强化都会产生一层薄的、却是十分有效的压应力表面层。可见看出,这些工艺方法使工件外层产生加工硬化,从而产生压应力,同时在与之相邻的内层中产生较小的拉应力。由于在各处都很容易获得压应力层,因此这些工艺方法适用于交变载荷和应力在拉应力与压应力之间变化的回转类零件。在采用这些工艺方法时,必须认真地控制滚轮的压力和进给量、喷丸的大小和速度等。对于这些,在工程技术书籍和期刊中有大量资料可供参考。

在滚压加工中,采用一个或一组坚硬的和非常光滑的滚子使零件表面产生加工硬化。这种加工适用于各种平面、圆柱面和圆锥面(见图 39.1)。滚压加工通过去除划痕和刀痕来降低表面粗糙度,并且能够产生有益的表面残余压应力,提高疲劳寿命。采用滚压法加工螺栓和螺钉上的螺纹早就已经成为一种成形加工方法。这种方法不仅能够生成螺纹,而且还能通过螺纹根部附近的变形和晶粒流动以及所产生的残余压应力来增加螺纹的强度。

挤孔这种加工方法是迫使一个尺寸稍大的坚硬的碳化钨或 AISI 521000 钢球通过板、套筒或者管子上的孔,以获得最终尺寸和较低的表面粗糙度。孔的长度可以等于其直径的 0.05~10 倍。挤孔机通常由非熟练工人操作,适用于小型零件的大量生产。这种方法能够提高硬度,因而能提高耐磨性,并且在孔的周围产生残余压应力,这种残余应力通常是有利的,例如在滚子链的链条中。链条承受很高的脉动拉应力,而且在孔表面及其附近有应力集中。采用钢球挤孔可以产生压应力,这会降低工作中的有效拉应力,将断裂事故减到最低限度。

喷丸强化是通过采用机械方式引起工件表面屈服而施加预应力的最常用的方法。在圆形打击物的冲击下,表面发生变形并形成许多浅坑,由于这些浅坑有向外扩展的趋势而使表面受压。

喷丸强化时,以直径为 0.2~5mm 的小圆钢丸或者冷硬铸铁丸高速冲击钢制零件。受压层的深度可以达到 1.25mm,小于锤击强化时受压层的深度,大致与所用球丸的直径和速度成正比。所产生的残余应力大约等于应变强化区的屈服强度的一半。

除了零件的某些内部形状外,喷丸强化可以用于大多数金属和形状,而且成本最低,因此喷丸强化得到了广泛的应用。对于软金属,可以采用玻璃球。螺旋弹簧通常要进行喷丸硬化,可以使其在脉动载荷作用下的许用应力提高60%。强度增加的原因之一可能是由于消除了拉丝时产生的能够起减低强度作用的纵向擦痕所致。同样,进行喷丸强化可以改善粗加工表面和粗磨表面的质量,使其变得光滑,这种方法可能比采用切削加工或者磨削加工的方法进行最终加工更为经济。喷丸强化不能用于轴承和其他精度要求高的精密配合表面。对喷丸强化后的工件进行精磨可能会消除部分或者全部残余应力。可以采用喷丸机对在传送带或者转台上移动的中小尺寸零件进行自动和连续地喷丸强化。

激光冲击强化(也称激光喷丸)是一种新兴的表面强化技术。它利用激光产生的高能量的冲击波在金属的被冲击区产生残余压应力。这种方法所需要的激光能量密度大约为 100~300 J/cm²,其脉冲持续时间为 30 ns 左右。这种表面处理方法生成的残余压应力层的厚度通常为 1 mm。激光冲击强化已经被成功地和可靠地应用于喷气发动机风扇叶片和诸如钛合金,镍合金与钢等材料的表面处理,用来提高其抗疲劳、耐磨损和抗腐蚀能力。

第四十一课　可靠性对产品销售的影响

可靠性具有销售价值吗? 答案是既具有又不具有。它完全取决于你所从事的行业的类型。在航空航天工业中工作的人们会作出绝对肯定的答复。细想一下阿波罗太空飞行任务的技术复杂性,它所达到的可靠性是卓越非凡的。阿波罗宇航员百分之百的安全返回地球,是 20 世纪后半期中可靠性技术的成功事例。有人可能会问,这种成功是怎么取得的? 这完全是因为在这个项目的构思阶段,就已经把可靠性连同其他性能参数一起写入技术要求之中。因此,投标应该建立在能够满足技术要求中所规定的可靠性指标的基础上。

在全球性市场经济的环境中,利润是力求达到高度可靠性的动力。为了完成要求的销售指标,每个公司或机构都应该在规定其他工程性能参数和运行特性的同时,确定必须达到的可靠性水平。制造费用通常可以决定所采用的结构形式,产品的制造费用应该维持在能使产品的售价在市场上具有竞争力的水平上。对于一种新产品,在工程研制开发费用中必须包括使产品达到要求的可靠性水平的费用。将一种可靠性不合格的新产品投放到市场中,会使周密的销售计划变得毫无意义。在许多情况下,可以把研制与开发作为销售策略的一部分,从而成为新产品产销计划的一个重要组成部分。在产品说明书中写明初始可靠性也会对产品销售起到有利的保证作用。如果顾客相信产品故障是在进行了为验证产品可靠性而设计的切合实际的试验工作之后发生的,他的不满意程度通常就会减小很多。

汽车制造业、航空发动机制造业和通风机制造业可以作为将研制开发与市场营销策略相结合的实例。当一种新型轿车面市后,在有可能购买这种汽车的顾客们的头脑中会产生两种互相对立的想法:

(1)它是新型的(因此比前一种型号"更好")。

(2)它是新型的(因此没有经过验证,在初期可能会出现许多故障,因而不能可靠地工作)。

应该采取的策略就是要使顾客相信,它既是新型的(所以比前一种型号更好,因而是能令人满意的),又是经过试验验证的(所以潜在的故障少,不需要经常去维修站)。为了能够做到这一点,在宣传材料和广告中用一些通俗易懂的试验来证明新型轿车的可靠性是非常重要的。例如,"这种轿车在推向市场之前已经行驶了十万英里",或者"200名经过挑选的驾驶员对这种轿车的试验表明,我们的新型轿车完全能够承担日常商业行驶任务"。在过去,这两个例子被用来告诉广大汽车驾驶人员,为了他们的利益,厂方已经进行了大量的研制开发工作,生产出令人满意和可靠的产品。

飞机发动机制造业的策略的类似之处在于,必须对航空公司或军队方面客户所提出的问题作出令人信服的答复。他们会说,他还记得前一种发动机在刚开始使用时出现的那些故障,你们采取了哪些措施来保证这种发动机不再出现类似的初期故障?使研制开发工作能够被接受的最简单方法就是去说明,前一种型号的发动机在交给航空公司或军队使用前在试验台上进行了10000小时的运行试验,但是新型发动机在交付使用前将在试验台上进行14000小时试验。在这种情况下,要最大限度在运用市场营销智慧,来确定顾客真正需要什么?顾客真正准备购买什么?

通风机行业的顾客所关心的问题,按通常的主次顺序为:价格、性能和可靠性。因此,资金中用于进行复杂试验的费用就比较少。对工程师来说,每一项试验都必须是有意义的和有说服力的,并且能使顾客相信,他安装这台设备之后就不需要费心维护。现场试验由于比较便宜,因而是非常重要的,但花费的时间比较长。在这种试验中,顾客可以看到设备的安装,安装细节和新产品令人满意的工作情况,这会使他对这种产品的性能产生信心。在现场试验中,潮湿、蒸汽或者脏的安装特别适合于进行"超载"试验。

如果某种通风机是用于农业环境的,在进行通风机的现场试验时,重要的是不要找一位精心、认真的用户,而是要找一位能够做出用水冲洗设备的方式来进行清洁的这种不符合操作规程的事的用户。用水冲洗这件事能够充分表明,电动机的轴上需要装有密封装置,以阻止水进入电动机而毁坏设备的工作可靠性。

第四十三课　质量与检测

根据美国质量管理协会的定义,质量是产品或服务能够满足规定需求而具有的特性和特征的总和。这个定义表明,应该首先确定顾客的需求,因为满足这些需求是达到质量目标的底线。然后应该把顾客的需求转化为产品的特性和特征,并据此进行设计工作和制订产品的技术要求。

除了正确地理解质量这个术语之外,对质量管理、质量保证和质量控制等术语的理解也是非常重要的。

质量管理是确定和实施质量方针的全部管理功能。质量管理是高层管理人员的职责。这项工作包括战略规划、资源分配和与之有关的质量规划活动。

质量保证包括能够给予满足规定需求的产品或服务以适当信任所必需的全部有计划,有组织的活动。这些活动的目标是给人们提供对质量体系能够正常工作的信任,而且包括对设计方案和技术要求妥善性的评价或者对生产作业能力的审查。内部质量保证的目标在于使公司的管理人员对产品质量具有信心,而外部质量保证的目的在于向公司产品的用户们在产品质量方面做出担保。

质量控制包括了为保持产品或服务的质量,使其能满足规定需求所采取的作业技术和有关活动。质量控制功能与产品的关系最为密切,它采用各种技术和活动来监视加工过程,并且消除生成不合格产品的根源。

许多过去的质量体系的设计目标是,在各个加工阶段中,将合格的产品与不合格的产品区分开来。那些被判定为不合格的产品必须经过重新加工以满足技术要求。如果它们不能被重新加工,就应该被当成废料处理。这种体系被称为"检测—改正"体系。在这种体系中,只有在对产品进行检测时,或者当用户使用产品时,才会发现其中存在的问题。由于检查人员本身固有的因素所决定,这种分类拣选工作的有效性经常低于90%。目前,预防型质量体系得到了越来越广泛的应用。在这种体系中,为了避免发生质量问题,首先强调对产品生产的各个阶段进行合理的规划和制订预防发生问题的措施。

对于一种产品满足需求或期望值的好坏程度的最终评价是由顾客和用户们作出的,而且会受到可以向这些顾客和用户出售同类产品的竞争对手的影响。这个最终评价是基于产品在整个使用寿命周期内的表现作出的,而不是仅仅只根据其在购买时的性能作出的。认识到这一点是很重要的。

如前所述,了解顾客的需求和期望是很重要的。此外,将一个企业中的全体员工的注意力集中在顾客和用户以及他们的需求上，会产生一个更有效的质量体系。例如，在对产品设计方案和技术要求进行小组讨论时，要对顾客应该得到满足的需求进行专门的讨论。

管理人员的一个基本职责是不懈地追求质量改善。应该认识到在当今的市场中，质量是一个由顾客们对产品日益增高的期望所驱动的动态目标，质量行动必须落实到公司每天的工作中。传统的作法是对某一产品确定一个适当的质量等级，然后集中全部力量使产品满足这个质量等级。这种方法不适合于在长期的生产中使用。与之相反，管理人员应该给企业确定这样的方向：一旦某种产品达到了适当的质量等级之后，还应对其进行不断的改进，使其达到更高的质量等级。

为了获得最有效的改进成果，管理人员应该明白质量与成本是互补的，而不是相互矛盾的。传统上，对管理人员的建议是必须在质量与成本之间作出选择。这是因为质量好的产品，其成本必然要比较高，而且其生产难度也比较大。目前，全世界范围内的经验已经表明，这是不正确的。高质量的产品从根本上实现了资源的优化利用，从而意味着高生产能力和低质量成本。此外，被顾客们认为质量高和在使用过程中性能可靠的产品，其销售量和市场覆盖率都会明显地高于同类产品。

质量成本的四种基本类型如下所述：

1. 预防成本是规划、实施和维持一个质量体系，使其能在经济的水平上保证产品性能满足质量要求的费用。进行统计过程控制方面知识的培训可以作为质量预防成本的一个例子。

2. 评价成本是确定产品性能与质量要求符合程度的费用。检验可以作为质量评价成本的一个例子。

3. 内部缺陷处置成本是在将产品、元件和材料的所有权转移给顾客之前就不能满足质量要求的费用。废品可以作为质量内部缺陷处置成本的一个例子。

4. 外部缺陷处置成本是在将所有权转移给顾客之后，产品不能满足质量要求的费用。根据产品保证书进行索赔可以作为外部缺陷处置成本的一个例子。

尽管确定质量控制等级的大部分工作是采用概率论和统计计算进行的，应用适当的和精确的数据采集过程为这项工作提供依据也是非常重要的。提供错误的数据会使最好的统计工作程序变得毫无意义。与机械加工过程类似，采集检验数据本身也是一个具有精确性、准确度、分辨率和可重复性限度的过程。

所有的检验和(或)测量过程都可以依据它们的精度和可重复性来确定范围，这与采用精度和可重复性来评价制造过程是一样的。可以进行受控制的对照实验，获得统计度量结果，用来确定对被测零件所采用的检测方法的性能和效果。对于每种方法适宜性的评价是基于在规定的检验工作中，应用这种方法所产生的标准偏差或置信水平作出的。

第四十五课　计算机在制造业中的应用

　　计算机正在将制造业带入信息时代。计算机长期以来在商业和管理方面得到了广泛的应用,它正在作为一种新的工具进入到工厂中,而且它如同蒸汽机在200多年前使制造业发生改变那样,正在使制造业发生着变革。

　　尽管基本的金属切削过程不太可能发生根本性的改变,但是它们的组织形式和控制方式必将发生改变。

　　从某一方面可以说,制造业正在完成一个循环。最初的制造业是家庭手工业:设计者本身也是制造者,产品的构思与加工由同一人完成。后来,形成了零件的互换性这个概念,生产被依照专业功能分割开来,可以成批地生产数以千计的相同零件。

　　今天,尽管设计者与制造者不再可能是同一个人,但在向集成制造系统前进的途中,这两种功能已经越来越靠近了。

　　可能具有讽刺意味的是,在市场需求高度多样化的产品的同时,制造业必须提高生产率和降低成本。消费者要求用较少的钱去购买高质量和多样化的产品。

　　计算机是满足这些要求的关键因素。它是能够提供快速反应能力、柔性和来满足多样化市场的唯一工具。而且,它是实现制造系统集成所需要的、能够进行详细分析和利用精确数据的唯一工具。

　　在将来计算机可能会是一个企业生存的基本条件,许多现今的企业将会被生产能力更高的企业组合所取代。这些生产能力更高的企业组合是一些具有非常高的质量、非常高的生产率的工厂。目标是设计和运行一个能够以高生产率的方式生产100%合格产品的工厂。

　　一个采用先进技术,充满竞争的世界正在要求制造业开始满足更多的需求,并且促使其本身采用先进的技术。为了适应竞争,一个公司会满足某些多少有点相互矛盾的需求,比如说,既要生产多样化和高质量的产品,又要提高生产率和降低价格。

　　在努力满足这些需求的过程中,公司需要一个采用先进技术的工具,一个能够对顾客的需求做出快速反应,而且能够从制造资源中获得最大收益的工具。

　　计算机就是这个工具。

　　成为一个具有"非常高的质量、非常高的生产率"的工厂,需要对一个非常复杂的系统进行集成。这只有通过采用计算机对制造业的所有组成部分——设计、加工、装配、质量保证、管理和物料输送进行集成才能够完成。

　　例如,在产品设计期间,交互式计算机辅助设计系统使得完成绘图和分析工作所需要的时间比原来减少了几倍,而且精确程度得到了很大的提高。此外,样机的试验与评价程序进一步加快了设计过程。

在制订制造计划时,计算机辅助工艺过程设计可以从数以千计的工序和加工过程中选择最好的加工方案。

在车间里,分布式智能以微处理器这种形式来控制机床、操纵自动装卸料设备和收集关于当前车间状态的信息。

但是这些各自独立的改革还远远不够。我们所需要的是由一个共同的软件从始端到终端进行控制的全部自动化的系统。

一般来说,计算机集成可以提供广泛的、及时的和精确的信息,可以改进各部门之间的交流与沟通,实施更严格的控制,通常能够提高整个系统的整体质量和效率。

例如,改进交流和沟通意味着会使设计具有更好的可制造性。数控编程人员和工艺装备设计人员有机会向产品设计人员提出意见,反之亦然。

因而可以减少技术方面的变更,而对于那些必要的变更,可以更有效地进行处理。计算机不仅能够更快地对变更之处作出详细说明,而且还能能把变更之后的数据告诉随后的使用者。

利用及时更新的生产控制数据可以制订更好的工艺规程和更有效率的生产进度。因而,可以使昂贵的设备得到更好的利用,提高零件在生产过程中的运送效率,减少在制品的成本。

产品质量也可以得到改进。例如,不仅可以提高设计精度,还可以使质量保证部门利用设计数据,避免由于误解而产生错误。

可以使人们更好地完成他们的工作。通过避免冗长的计算和书写工作——这还不算查找资料所浪费的时间——计算机不仅使人们更有效地工作,而且还能把他们解放出来去做只有人类才能做工作:创造性地思考。

计算机集成还会吸引新的人才进入制造业。人才被吸引过来的原因是他们希望到一个现代化的、技术先进的环境中工作。

在制造工程中,CAD/CAM减少了工艺装备设计、数控编程和进行工艺过程设计所需要的时间。而且,在同时加快了响应速度,这最终会使目前外委加工的工作由公司内部人员来完成。

第四十七课　计算机辅助工艺设计

根据《工具与制造工程师手册》,工艺设计是系统地确定一套能够经济地和有竞争力地将产品制造出来的方法。它主要由选择、计算和建立工艺文件组成。对加工方法、机床、刀具、工序和顺序必须要进行选择。对于一些参数如进给量、速

2. 激光切割。

3. 电子束焊接。

数字控制还使得机床比它们采用人工操纵的前辈们的用途更为广泛。一台数控机床可以自动生产很多种类的零件,每一个零件都可以有不同的和复杂的加工过程。数控可以使生产厂家承担那些对于采用人工控制的机床和工艺来说,在经济上是不划算的产品的生产任务。

同许多先进技术一样,数控诞生于麻省理工学院的实验室中。数控这个概念是50年代初在美国空军的资助下提出来的。

APT(自动编程工具)语言是1956年在麻省理工学院的伺服机构实验室中被设计出来的。这是一个专门适用于数控的编程语言,使用类似于英语的语句来定义零件的几何形状,描述切削刀具的形状和规定必要的运动。APT语言的研究和发展是在数控技术进一步发展过程中的一大进步。最初的数控系统与今天应用的数控系统是有很大差别的。在那时的机床中,只有硬线逻辑电路。指令程序写在穿孔纸带上,如图49.1所示,它后来被塑料磁带所取代。采用带阅读机将写在纸带或磁带上的指令给机器翻译出来。所有这些共同构成了机床数字控制方面的巨大进步。然而,在数控发展的这个阶段中还存在着许多问题。

一个主要问题是穿孔纸带的易损坏性。在机械加工过程中,载有编程指令信息的纸带断裂和被撕坏是常见的事情。在机床上每加工一个零件,都需要将载有编程指令的纸带放入阅读机中重新运行一次。因此,这个问题变得很严重。如果需要制造100个某种零件,则应该将纸带分别通过阅读机100次。易损坏的纸带显然不能承受严酷的车间环境和这种重复使用。

这就导致了一种专门的塑料磁带的研制。在纸带上通过采用一系列的小孔来载有编程指令,而在塑料带上通过采用一系列的磁点来载有编程指令。塑料带的强度比纸带的强度要高很多,这就可以解决常见的撕坏和断裂问题。然而,它仍然存在着两个问题。

其中最重要的一个问题是,对输入到带中的指令进行修改是非常困难的,或者是根本不可能的。即使对指令程序进行最微小的调整,也必须中断加工,制作一条新带。而且带通过阅读机的次数还必须与需要加工的零件的个数相同。幸运的是,计算机技术的实际应用很快解决了数控技术中与穿孔纸带和塑料带有关的问题。

在形成了直接数字控制(DNC)这个概念之后,可以不再采用纸带或塑料带作为编程指令的载体,这样就解决了与之相关的问题。在直接数字控制中,几台机床通过数据传输线路连接到一台主计算机上。操纵这些机床所需要的程序都存储在这台主计算机中。当需要时,通过数据传输线路提供给每台机床。直接数字控制是在穿孔纸带和塑料带基础上的一大进步。然而,它也有着同其他依赖于主计算机的技术一样的局限性。当主计算机出现故障时,由其控制的所有机床都将停止工作。这个问题促使了计算机数字控制技术的产生。

微处理器的发展为可编程逻辑控制器和微型计算机的发展做好了准备。这两种技术为计算机数控(CNC)的发展打下了基础。采用 CNC 技术后,每台机床上都有一个可编程逻辑控制器或者微机对其进行数字控制。这可以使得程序被输入和存储在每台机床内部。它还可以在机床以外编制程序,并将其下载到每台机床中。计算机数控解决了主计算机发生故障所带来的问题,但是它产生了另一个被称为数据管理的问题。同一个程序可能要分别装入十个相互之间没有通讯联系的微机中。这个问题目前正在解决之中,它是通过采用局部区域网络将各个微机连接起来,以利于更好地进行数据管理。

第五十一课　培训编程人员

熟练的零件编程人员是有效地利用数控机床的基本要求。他们的工作决定了这些机床的工作效率和在机床本身,工厂的数控辅助设备和管理费用等方面的投资所能得到的经济回报。

目前,熟练的零件加工数控编程人员非常短缺。这不仅表明了在机械加工业普遍缺少有经验的人员,而且也表明随着越来越多地通过应用数控机床来增加生产能力、通用性和生产率,对编程人员的需求也日益增多。

就一个行业而言,明显的答案是通过培训来培养新的编程人员,而且可以通过许多途径进行这种培训。首先应该确定编程人员应该具备什么条件和参加培训的编程人员应该学习什么?

根据全国机床制造厂商协会编写的《选择适当的数控编程方法》这本小册子,手工编程人员主要应该具备下列各项条件:

机械制造经验　编程人员对其要进行编程的数控机床的性能应该有透彻的了解,还要了解车间中其他机床的基本性能。他们还应在金属切削原理和实践、刀具的切削能力、夹具和夹持技术等方面有广泛的知识和感受力。编程人员还应当在能够显著降低生产成本的制造工程技术方面得到适当的培训。

空间想象力　编程人员应该能够想象出零件的三维形状,机床的切削运动,在刀具、工件、夹具或者机床本身之间可能产生的干涉。

数学　算术、代数、三角学、几何等方面的知识是非常重要的。高等数学,诸如高等代数、微积分等,通常是不需要的。

对细节的关注　编程人员应该是具有敏锐的观察能力并且是非常认真仔细精确的人。要改正在机床编排过程中发现的编程错误,可能会花费很多金钱和时间。

在这本小册子的另一个地方提到:"与计算机辅助编程相比,手工编程要求编程人员在机床及其控制系统、加工过程、计算方法等方面掌握更多的知识。另一方面,采用计算机辅助编程,应该掌握计算机编程语言和运用这种语言所需要的计算机系统知识。一般说来,由于涉及许多细节知识,手工编程更慢和更为复杂。在计算机辅助编程中,这些细节知识都包含在计算机系统(处理程序,后置处理程序)中。"

数控技术和培训方面的专家通常同意这些条件和要求,他们还增加了一些次要方面的细节要求。诸如:阅读图纸、不同种类金属的切削性能、车间测量仪器的使用、公差、安全措施等方面的知识。

应该到哪里去寻找可以进行培训的学员呢? 首先在你自己工厂的车间里。埃德华·F·斯罗斯,一位辛辛那提·米拉克龙公司主管销售的副总经理,这样说:"我们在将优秀的车工和铣工培训成编程人员方面有很成功的经验。他们没有认识到,但是实际上他们工作的大部分时间都在编程,而且他们有车间所需要的基础数学和三角学知识。你可以很容易地教会他们编程。反之,将一个能力很强的数学家培养成编程人员则很困难。对他来说,轨迹的编程很容易。但是如何完成轨迹的加工——进给量、切削速度等——这可能需要更多的培训。"

采用功能较强的计算机辅助编程后,对编程人员在金属切削知识方面的要求就降低了。通过应用这种软件,辛辛那提·米拉克龙公司成功地雇用了一些大学刚毕业的学生,其中还包括一些非技术专业毕业的大学生。通过先对这些学员在工厂中进行机床实际操作培训,然后再对他们进行编程方面的培训,将他们培养成零件数控加工的编程人员。

当然,所有的数控机床生产厂家对他们的产品会提供某种编程培训,而且大部分厂家提供正式的培训计划。例如,米拉克龙公司的销售部门就有 20 名专门培训客户的教师。这个公司对参加培训的学员的预先要求有下列各项:

"参加人员应该了解机械加工车间的安全规程,能够看懂零件图、剖面图和工件的数控程序单。

"需要掌握平面几何、直角三角学和公差的基本知识。

"还需要具有零件手工数控编程、数控机床的编排和操作过程、零件加工、金属切削技术、刀具和夹具等方面的知识。"

将具有以上基础的人送到学校,将能保证数控机床的使用者从他们所花在培训上的钱(尽管培训费已经包括在购买机床的基本费用中,接受培训的人员需要花费一周的时间、交通费、生活费)中,得到最大的收益。

第五十三课　工业机器人

有许多关于工业机器人的定义。采用不同的定义,全世界各地工业机器人装置的数量就会发生很大的变化。在制造工厂中使用的许多单用途机器可能会看起来像机器人。这些机器只具有单一的功能,不能通过重复编程的方式去完成不同的工作。这种单一用途的机器不能满足被人们日益广泛接受的关于工业机器人的定义。

国际标准化组织(ISO)对工业机器人的定义为:一种能够自动控制的,可重复编程的多功能操作机,它可以是固定式的也可以是移动式的,应用于工业自动化领域。

其他的一些协会,例如美国机器人协会(RIA),英国机器人协会(BRA)等都对工业机器人提出了他们各自的定义。由美国机器人协会提出的定义为:

机器人是一种用于移动材料、零件、工具或者专用装置的,通过可编程序动作来执行多种任务并具有可重复编程能力的多功能操作机。

在所有的这些定义中有两个共同点。它们是"可重复编程"和"多功能"这两个词。正是这两个特征将真正的机器人与现代制造工厂中使用的各种单一用途的机器区分开来。

"可重复编程"这个词意味着两件事:机器人根据编写的程序工作,以及可以通过重新编写程序来使其适应不同种类的制造工作的需要。

"多功能"这个词意味着机器人能通过重复编程和使用不同的末端执行器,来完成不同的制造工作。围绕着这两个关键特征所撰写的定义已经变成了被制造业的专业人员所接受的定义。

第一个关节式手臂于1951年被研制出来,供美国原子能委员会使用。在1954年,第一个可以编程的机器人由乔治 C. 德沃尔设计出来。它是一个不复杂的,可以编程的物料搬运机器人。

第一个商业化生产的机器人在1959年研制成功。通用汽车公司在1962年安装了第一个用于生产线上的工业机器人。它在位于美国新泽西州的一家汽车厂中被用来从压铸机中取出红热的车门拉手以及诸如此类的汽车零件。它最显著的特点是通过采用手爪,避免由人去接触那些刚刚由熔化的金属形成的汽车零件。它有五个自由度。它是由万能自动化公司(Unimation)生产的。

在1973年,辛辛那提·米拉克龙(Milacron)公司研制出 T^3 工业机器人,在机器人的控制方面取得了重大的进展。T^3 机器人是第一个商业化生产的采用小型计算机控制的机器人。T^3 机器人和它的所有运动如图 53.1 所示,它也可以被称为关节式球面机器人。

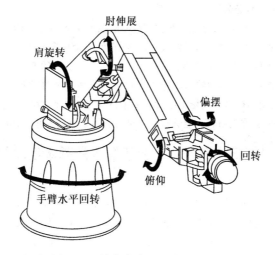

图 53.1　关节式球面机器人

　　从那时起,机器人技术在很多方面都得到了发展,这包括焊接、喷漆、装配、机床上下料和检测。

　　在过去的三十年中,机器人在许多汽车制造厂中占据了主要地位。在一个工厂中,通常有数以百计的工业机器人工作在全自动生产线上。例如,在一条自动生产线上,车辆底盘装在输送机上,在通过一连串的机器人工作站时进行诸如焊接、喷漆和最后的装配等项工作。

　　在印刷电路板的大批量生产过程中,装配工作几乎完全都是采用抓-放型机器人进行的。通常采用平面关节型装配机器人(SCARA),它可以抓取微小的电子元器件并以非常高的精度将其放到印刷电路板上。这类机器人每小时可以放置成千上万个元器件,其速度、精度和可靠性都远远超过了人类。

　　工业机器人成本的降低是促进它们的使用量增长的一个主要原因。从 20 世纪 70 年代开始,工资的快速增长大大增加了制造业中的人工费用。为了生存,制造厂家被迫考虑采用任何能够提高生产率的技术。为了在全球性市场经济的环境中具有竞争能力,制造厂家必须以比较低的成本,生产出质量更好的产品。其他的因素,诸如寻找更好的方式来完成带有危险性的制造工作,也促进了工业机器人的发展。但是,其根本原因一直是,而且现在仍然是提高生产率。

　　机器人的主要优点之一是它们可以在对于人类来说是危险的环境中工作。采用机器人进行焊接和切断工作是比由人工来完成这些工作更为安全的例子。大部分现代机器人被设计用在对人类来说是不安全和非常困难的环境中工作。例如,可以设计一个机器人来搬运非常热或非常冷的物体,这些物体如果用人工搬运,则存在不安全因素。

尽管机器人与工作地点的安全密切相关,它们本身也可能是危险的。应该仔细地设计和配置机器人和机器人单元,使它们不会伤害人类和其他机器。应该精确地计算出机器人的工作空间,并且在这个工作空间的四周清楚地标出危险区域。可以通过设置障碍物来阻止工人进入机器人的工作空间。即使有了这些预防措施,在使用机器人的场地中设置一个自动停止工作的系统仍然不失为一个好主意。这个系统应当具有能够检测出是否有需要自动停止工作的要求的能力。

第五十五课　工业机器人的基本组成部分

为了评价机器人各个组成部分的功能和性能,我们可以同时观察在从货架上抓取物体时,使用手工工具时或者是操纵机器时,我们的手臂,手腕,手和手指的各种运动的灵活性和能力。工业机器人的基本组成如下所述:

操作机　操作机是一个能够提供类似人的手臂和手腕运动的机械装置。如同图54.1中所示的机器人,操作机通常由构件(它们的功能类似于人体中的骨头)和关节(也被称为"运动副")以串联的方式连接而成。对于一个典型的六自由度机器人(如图53.1所示),前三个构件和关节构成了手臂,而后三个关节构成了手腕。手臂的功能是将一个物体放在三维空间中的某一位置,手腕确定此物体的方位。

末端执行器　末端执行器是安装在机器人的操作机端部的装置。它相当于人的手。根据其用途,常见的末端执行器有以下几种:

(1) 用于物料搬运的手爪,电磁铁和真空吸盘;

(2) 用于喷涂的喷枪;

(3) 进行点焊和弧焊的焊接装置;

(4) 诸如电钻等类的电动工具;和

(5) 测量仪器。

末端执行器通常是按照客户的特殊要求定做的。最常用的末端执行器是机械式手爪,它们有二个或多个手指。具有二个手指的手爪(图55.1)只能抓取形状简单的物体,具有多个手指的手爪则能够完成更为复杂工作。为某一特定的用途选择末端执行器时,应该考虑诸如承载能力,环境,可靠性和成本等项因素。

驱动器　驱动器就像机器人的肌肉,它们为操作机和末端执行器提供运动。根据其工作原理,可以将它们分为气动、液压和电气驱动器。

气动驱动器利用活塞或气动马达将由压缩机提供的压缩空气转换为机械能。气动驱动器只有极少数运动部件,这决定了它们具有较高的可靠性和只需要较低的维修费用。在所有的驱动器中,它是价格最便宜的。但是,由于空气的可压缩性,气动驱动器不适合应用于搬运重物并且需要精确控制的场合。

液压驱动器利用油等高压流体将力传到需要的作用点。液压驱动器在外观上很像气动驱动器。二者相比,液压驱动器的工作压力要高很多(通常为 7～17MPa)。它们适合用于需要大功率的场合。与电气驱动的机器人相比,液压驱动的机器人具有更强的承受冲击载荷的能力。

作为操作机的驱动器,电动机的使用数量最多。直流(DC)电动机具有很高的扭矩/体积比。它们还具有高精度、高加速度和高可靠性。尽管它们的功率重量比不如气动驱动器和液压驱动器,其可控性使得它们适用于中小型操作机。

交流(AC)电动机和步进电机在工业机器人中并不常用。前者在控制方面的困难和后者只能产生较低的扭矩都限制了它们的应用。

传感器 传感器可以将一种信号转换为另一种信号。例如,人眼可以将光图像转换为电信号。传感器可以分为几种类型:视觉、触觉、位置、力觉、速度、加速度等。其中的一些传感器将在56课中作详细说明。

数字控制器 数字控制器是一种专用的电子装置,它有中央处理器(CPU)、存储器,有时还会有硬盘。在机器人系统中,这些元器件被装在一个叫做控制器的密封盒子里。它被用来控制操作机和末端执行器的运动。由于计算机具有与数字控制器相同的特性,它也可以被用来作为机器人的控制器。

模数转换器 模数转换器(简称 ADC)是一个将模拟信号转换为数字信号的电子器件。这种电子器件与传感器和机器人控制器相连接。通常,模数转换器将输入的模拟电压(或电流)转换成与电压或电流成正比的数值。例如,模数转换器(ADC)可以将由于应变片的应变而产生的电压信号转换为数字信号,使得机器人的数字控制器能够处理这个信息。

数模转换器(DAC) 数模转换器(DAC)把从机器人控制器中获得的数字信号转换为模拟信号,以启动驱动器。为了使驱动器(例如,一台直流电动机)工作,与数字控制器相连接的数模转换器(DAC)要把它的数字信号转换回模拟信号,也就是,直流电动机的电压。

放大器 一般说来,放大器是一种能够改变,通常是增大信号幅值的装置。由数字控制器中发出的指令通过数模转换器后生成的模拟信号很弱,只有将它们放大之后才能驱动机器人中的电动机。

第五十七课 工程师在制造业中的作用

许多工程师的职责是进行产品设计,而产品是通过对材料的加工制造而生产出来的。设计工程师在选择材料、制造方法等方面起着关键的作用。一个设计工程师应该比其他的人更清楚地知道他的设计需要达到什么目的。他知道他对使用荷载和使用要求所做的假设,产品的使用环境,产品应该具有的外观形貌。为了满足这些要求,他必须选择和规定所使用的材料。通常,为了利用材料并使产品具有所期望的形状,设计工程师知道应该采用哪些制造方法。在许多情况下,选择了某种特定材料就可能意味着已经确定了某种必须采用的加工方法。同时,当决定采用某种加工方法后,很可能需要对设计进行修改,以使这种加工方法能够被有效而经济地应用。某些尺寸公差可以决定产品的加工方法。

总之,在将设计转变为产品的过程中,必须有人做出这些决定。在大多数情况下,如果设计人员在材料和加工方法方面具有足够的知识,他会在设计阶段作出最为合理的决定。否则,作出的决定可能会降低产品的性能,或者使产品变得过于昂贵。显然,设计工程师是制造过程中的关键人物,如果他们能够进行面向生产(即可以进行高效率生产)的设计,就会给公司带来效益。

制造工程师们选择和调整所采用的加工方法和设备,或者监督和管理这些加工方法和设备的使用。一些工程师进行专用工艺装备的设计,以使通用机床能够被用来生产特定的产品。这些工程师们在机床、加工能力和材料方面必须具有广泛的知识,以使机器在没有过载和损坏,而且对被加工材料没有不良影响的情况下,更为有效地完成所需要的加工工序。这些制造工程师们在制造业中也起到重要作用。

少数工程师们设计在制造业中使用的机床和设备。显然,他们是设计工程师。而且对于他们的产品而言,他们同样关心设计、材料和制造方法之间的相互关系。然而,他们更多地关心他们所设计的机床将要加工的材料的性能和机床与材料之间的相互作用。

还有另外一些工程师——材料工程师,他们致力于研制新型的和更好的材料,他们也应该关心这些材料的加工方法和加工对材料性能的影响。

尽管工程师们所起的作用可能会有很大差别,但是,大部分工程师们都必须考虑材料与制造工艺之间的相互关系。

低成本制造并不是自动产生的。在产品设计、材料选择、加工方法和设备的选择,工艺装备选择和设计之间都有着非常密切的相互依赖关系。这些步骤中的每一个都必须在开始制造前仔细地加以考虑、规划和协调。这种从产品设计到实际生产的准备工作,特别是对于复杂产品,可能需要数月甚至数年的时间,并且可能花费很多钱。典型的例子有,对于一种全新的汽车,从设计到投产所需要的时间大约为2年,而一种现代化飞机则可能需要4年。

随着计算机和由计算机控制的机器的出现,我们进入了一个生产计划的新时代。采用计算机将产品的设计功能与制造功能集成,被称为CAD/CAM(计算机辅助设计/计算机辅助制造)。这种设计被用来制定加工工艺规程和提供加工过程本身的编程信息。可以根据供设计与制造用的中心数据库内的信息绘制零件图,需要时可以生成加工这些零件时所使用的程序。此外,对加工后零件的计算机辅助试验与检测也得到了广泛的应用。随着计算机价格的降低和性能的提高,这种趋势将毫无疑问地得到不断加速的发展。

第五十九课　信息时代的机械工程

在 80 年代初期,工程师们曾经认为要加快产品的研制开发,必须进行大量的研究工作。结果是实际上只进行了较少的研究工作,这是因为产品开发周期的缩短,促使工程师们尽可能地利用现有的技术。研制开发一种创新性的技术并将其应用在新产品上,是有风险的,并且易于招致失败。在产品开发过程中采用较少的步骤是一种安全的和易于成功的方法。

对于资金和人力都处于全球性环境中的工程界而言,缩短产品研制开发周期也是有益的。能够设计和制造各种产品的人可以在世界各地找到。但是,具有创新思想的人则比较难找。对于你已经进行了 6 个月的研制开发工作,地理上的距离已经不再是其他人发现它的障碍。如果你的研制周期较短,只要你仍然保持领先,这种情况并不会造成严重后果。但是如果你正处于一个长达 6 年的研制开发过程的中期,一个竞争对手了解到你的研究工作的一些信息,这个项目将面临比较大的麻烦。

工程师们在解决任何问题时都需要进行新的设计这种观念很快就过时了。在现代设计中的第一步是浏览因特网或者其他信息系统,看其他人是否已经设计了一种类似于你所需要的产品,诸如传动装置或者换热器等。通过这些信息系统,你可能发现有些人已经有了制造图纸,数控程序和制造你的产品所需要的其他所有东西。这样,工程师们就可以把他们的职业技能集中在尚未解决的问题上。

在解决这类问题时,利用计算机和计算机网络可以大大增强工程小组的能力和效率。这些信息时代的工具可以使工程小组利用大规模的数据库。数据库中有材料性能、标准、技术和成功的设计方案等信息。这些经过验证的设计可以通过下载直接应用,或者通过对其进行快速、简单的改进来满足特定的要求。将产品的技术要求通过网络送出去的远程制造也是可行的。你可以建立一个没有任何加工设备的虚拟公司。你可以指示制造商,在产品加工完成后,将其直接送给你的客户。定期访问你的客户可以保证你设计的产品按照设计要求进行工作。尽管这些研制开发方式不可能对每个公司都完全适用,但是这种可能性是存在的。

过去客户设计的产品通常是由小公司来制造。大公司不屑于制造这种产品——它们讨厌瞄准机会的市场,或者是与客户设计的小批量产品打交道。"这就是我的产品",一家大公司这样说:"这是我们能够制造出来的最好产品——你应该喜欢它。如果你不喜欢,在这条街上有一家小公司,它会按你的要求去做"。

今天,因为顾客们有较大的选择余地,几乎所有的市场都是瞄准机会的市场。如果你不能使你的产品满足某些特定客户的要求,你将失掉你的市场份额中的一大部分,或者失掉全部份额。由于这些瞄准机会的市场是经常变化的,你的公司应该对市场的变化作出快速的反应。

瞄准机会的市场和根据客户要求进行设计这种现象的出现改变了工程师们进行研究工作的方式。今天,研究工作通常是针对解决特定问题进行的。现在许多由政府资助或者由大公司出资开发的技术可以在非常低的成本下被自由使用,尽管这种情况可能是暂时的。在对这些技术进行适当改进后,它们通常能够被直接用于产品开发,这使得许多公司可以节省昂贵的研究经费。在主要的技术障碍被克服后,研究工作应该主要致力于产品的商品化方面,而不是开发新的,有趣的,不确定的替换产品。

采用上述观点看问题,工程研究应该致力于消除将已知技术快速商品化的障碍。工作的重点是产品的质量和可靠性,这些在当今的顾客的头脑中是最重要的。很明显,一个质量差的声誉是一个不好的企业的同义词。企业应该尽最大的努力来保证顾客得到合格的产品,这个努力包括在生产线的终端对产品进行严格的检验和自动更换有缺陷的产品。

研究工作应该着重考虑诸如可靠性等因素带来的成本效益。当可靠性提高时,制造成本和系统的最终成本将会降低。如果在生产线的终端产生了30%的废品,这不仅会浪费金钱,也会给你的竞争对手创造一个利用你的想法制造产品,并将其销售给你的客户的良机。

提高可靠性和降低成本这个过程的关键是深入、广泛地利用设计软件。设计软件可以使工程师们加快每一阶段的设计工作。然而,仅仅缩短每一阶段的设计时间,可能不会显著地缩短整个设计过程的时间。因而,必须致力于采用并行工程软件,这样可以使所有设计组的成员都能使用共同的数据库。

随着我们步入信息时代,要取得成功,工程师们在技术开发和技术管理方面都应该具有一些独特的知识和经验。成功的工程师们不但应该具有宽广的知识和技能,而且还应该是某些关键技术或学科的专家,他们还应该在社会因素和经济因素对市场的影响方面有敏锐的洞察能力。将来,花在解决日常工程问题上的费用将会减少,工程师们将会在一些更富有挑战性,更亟待解决的问题上协同工作,大大缩短解决这些问题所需要的时间。计算机和网络使工程师们具有了越来越强的解决问题的能力,这也给他们的工作带来了很大的希望和喜悦。我们已经开始了工程实践的新阶段。机械工程是一个伟大的行业,当我们充分利用了信息时代所提供的机遇后,它将变得更加伟大。

第六十一课　如何撰写科技论文

题目　作者在准备论文题目时,应该记住一个明显的事实:论文的题目将被成千上万人读到。如果有能够完整地读完整篇论文的人的话,也可能只是少数几个人。大多数读者或者会通过原始期刊,或者会通过二次文献(文摘或索引)阅读到论文的题目。因此,题目中的每一个词都应该仔细地推敲,词与词之间的关系也应该细心处理。

对于可能阅读期刊目录中论文题目的读者来说,题目中每个词的含意和词序是很重要的。这对于所有可能使用文献的人,包括通过二次文献查找论文的人(可能是大多数)也是同样重要。因此,题目不仅仅作为论文的标记,它还应该适合于工程索引,科学引文索引等机器索引系统。大多数索引和摘要都采用"关键词"分类法。因此,在确定论文题目时,最重要的是作者应该提供能够正确表达文章内容的"关键词",也就是说文章的题目用词应该限于既容易理解,又便于检索,还能使文章的重要内容突出的那些词。

摘要　摘要应该是论文的缩写版本。它应该是论文各主要章节的简要总结。一篇写得好的摘要能使读者迅速而又准确地了解论文的基本内容,以决定他们是否对此论文感兴趣,进而决定他们是否要阅读全文。摘要一般不超过250个单词,并应该清楚地反映论文的内容。许多人会阅读原始期刊或工程索引或者另外一种其他二次出版刊物上刊登的摘要。

摘要应该:(1)阐述该项研究工作的主要目的和范围;(2)描述所使用的方法;(3)总结研究成果;(4)阐述主要结论。

摘要决不应该提及论文中没有涉及的内容或结论。在摘要中不要引用与该论文有关的参考文献(在极少的情况下除外,例如对以前发表过的方法的改进)。

引言　当然,论文的第一部分应该是引言。引言的目的是向读者提供足够的背景知识,使读者不需要阅读过去已经发表的有关此课题的论文,就能够了解和评价目前的研究成果。最重要的是,你应该简要地说明写这篇论文的目的。应该慎重地选择参考文献以提供最重要的背景资料。

实验过程　"实验过程"一节的主要目的是描述实验过程和提供足够的细节,以使有能力的研究人员可以重复这个实验。如果你的方法是新的(从未发表过的),那你就应该提供所需要的全部实验细节。然而,如果这个实验方法已经在正规的期刊上发表过,那么只要给出参考文献就可以了。

认真撰写这一节是非常重要的,因为科学方法的核心就是要求你的研究成果不仅有科学价值,而且也必须是能够重复的;为了判断研究成果能否重复,就必须为其他人提供进行重复实验的依据。不太可能重复的实验是不可取的;必须具有产生同样或相似结果的可能性,否则你的论文的科学价值就不大。

当你的论文受到同行们的审核时,一个好的审稿人会认真地阅读"实验过程"这一节。如果他确实怀疑你的实验能够被重复,不管你的研究成果多么令人敬畏,这个审稿人都会建议退回你的稿件。

结果 现在我们进入论文的核心部分——数据。论文的这部分被称为"结果"。"结果"一节通常由两部分组成。首先,你应该对实验作全面的叙述,提出一个"大的轮廓",但不要重复已经在"实验过程"一节中提到的实验细节。其次,你应该提供数据。

当然,这部分的写作不是一件很容易的事。你会如何提供数据呢?通常不能直接将实验笔记本上的数据抄到稿件上。最重要的是,在稿件中你应该提供有代表性的数据,而不是那些无限重复的数据。

"结果"一节应该写得清晰和简练,因为"结果"是由你提供给世界的新知识组成。论文的前几部分("引言"、"实验过程")告诉人们你为什么和如何得到这些结果的;而论文后面的部分("讨论")则告诉人们这些结果意味着什么。因此,很明显,整篇论文都是以"结果"为基础的。所以"结果"必须以确切而清晰的形式给出。

讨论 与其他章节相比,"讨论"一节所写的内容更难于确定。因此,它是最难写的一节。不管你知道与否,在许多论文中尽管数据正确,而且能够引起人们的兴趣,但是由于讨论部分写得不好也会遭到期刊编辑的拒绝。甚至更为可能的是,在"讨论"中所作的阐述使得数据的真正含义变得模糊不清,而使论文遭到退稿。

一个好的"讨论"章节的主要特征如下所述:

1.设法给出"结果"一节中的原理,相互关系,和归纳性解释。应该记住,一个好的"讨论"应该对"结果"进行讨论和论述,而不是扼要重述。

2.要指出任何的例外情况或相互关系中有问题的地方,并应该明确提出尚未解决的问题。决不要冒着很大的风险去采取另一方式,即试图对不适合的数据进行掩盖或捏造。

3.要说明和解释你的结果与以前发表过的研究结果有什么相符(或者不相符)的地方。

4.要大胆地论述你的研究工作的理论意义以及任何可能的实际应用。

5.要尽可能清晰地叙述你的结论。

6.对每一结论要简要叙述其论据。

在描述所观察的事物之间的相互关系时,你并不需要得出一个广泛的结论。你很少有能力去解释全部真理;通常,你尽最大努力所做的就是像探照灯那样照耀在真理的某一方面。你在这个方面的真理是靠你的数据来支持的;如果你将你的数据外延到更大的范围,那就会显得荒唐,这时甚至连你的数据所支持的结论也可能会受到怀疑。

Glossary

abatement [ə'beitmənt] *n*. 减少,减轻,降低,抑制,削弱

abrade [ə'breid] *v*. 擦掉,磨蚀,磨掉

abrasion [ə'breiʒən] *n*. 磨耗,磨损,磨损量

abrasive [ə'breisiv] *n*. 磨料,研磨材料,研磨剂;*a*. 磨料的,磨蚀的

abut [ə'bʌt] *v*. 邻接,毗连,支撑;*n*. 端,尽头,支架

acceleration [æk͵selə'reiʃən] *n*. 加速度,加速度值,促进,加快

accessibility [æk͵sesi'biliti] *n*. 可接近性,易维护性,可存取性

accessory [æk'sesəri] *a*. 附属的,附带的,次要的;*n*. 附件,配件,辅助装置

activate ['æktiveit] *v*.;*n*. 开动,启动,驱动,触发,促动

actuate ['æktjueit] *v*. 开动,驱动,使动作,操纵,驱使

actuator ['æktjueitə] *n*. 驱动器,执行机构

adverse environment 恶劣环境,不利的环境

AISI = American Iron and Steel Institute 美国钢铁学会

alignment [ə'lainmənt] *n*. 直线对准,调准,对中心

alphanumeric [͵ælfənju'merik] *a*. 字母数字;混合编制的;*n*. 字母数字符号

alternating current 交流,交流电

amplifer ['æmpli͵faiə] *n*. 放大器

amplification [͵æmplifi'keiʃən] *n*. 放大(系数,作用),增强,扩大

amplitude ['æmplitjuːd] *n*. 振幅,波幅,幅度,范围

analog ['ænələg] *n*. 模拟量,模拟装置,模拟系统

analogous [ə'næləgəs] *a*. 类似的,类比的,模拟的

angular ['æŋgjulə] *a*. 角的,角度的,倾斜的

anneal [ə'niːl] *v*.;*n*. 退火,(加热)缓冷,逐渐冷却

annular ['ænjulə] *a*. 环形的,环的

annular clearance 环状间隙

antiskid ['ænti'skid] *n*.;*a*. 防滑的,防滑轮胎纹

apparatus [æpə'reitəs] *n*. 装置,设备,器具,机器

arrangement [ə'reindʒmənt] *n.* 配置,布局,构造,方案,装置
articulated [ɑ:'tikjuleitid] *a.* 铰链的,有活关节的,关节式的
artificial intelligence 人工智能
artisan [ɑ:ti'zæn] *n.* 技工,工匠
asperity [æs'periti] *n.* (表面上的)粗糙,不平滑,凹凸不平
assembly drawing 装配图
assembly line 装配线
assn = association 协会,学会
assortment [ə'sɔ:tmənt] *n.* 种类,花色品种,分类,分级
attenuation [ə,tenju'eiʃən] *n.* 衰减(现象,量),减弱,降低
availability [ə,veilə'biliti] *n.* 可得到的,存在,可利用,使用价值
avert [ə'və:t] *v.* 防止,避免,避开
axial ['æksiəl] *a.* 轴的,轴向的
axis ['æksis] *n.* 轴线,轴
back engagement 背吃刀量
backlash ['bæklæʃ] *n.* 反向间隙,回程误差
backstroke ['bækstrəuk] *n.*;*v.* 返回行程,回程
ball screw 滚珠丝杠副
ballizing ['bɔ:laiziŋ] *n.* 挤孔,球推压法
bandwidth ['bændwidθ] *n.* 带宽,频带宽度
bar stock 棒料
basic size 基本尺寸,公称尺寸,规定尺寸
bearing ['bɛəriŋ] *n.* 轴承,支承,承载
bearing surface 承压面,支承面
bed-type milling machine 工作台不升降式铣床,床身式铣床
bench type 台式
bench-type drilling machine 台式钻床
bending ['bendiŋ] *n.* 弯曲,弯曲度,挠曲,挠曲度
bending moment diagram 弯矩图
beryllium [bə'riljəm] *n.* 铍
bevel ['bevəl] *n.* 斜角,倾斜;*v.* 斜切,斜截,削平
biaxial [bai'æksiəl] *a.* 二轴的,二维的

blade ['bleid] *n*. 叶片,桨片,刀片

blast [blɑ:st] *n*.; *v*. 爆炸,冲击,喷砂,喷丸

blind hole 盲孔

blueprint ['blu:'print] *n*.; *v*. (晒)蓝图,设计图,(订)计划

blur [blə:] *v*. 弄脏,变模糊,影像位移; *n*. 污迹,模糊

boom [bu:m] *n*. 吊杆,起重杆; *v*.; *n*. 繁荣,兴旺,畅销

bore [bɔ:] *n*. 枪膛,汽缸筒,孔,孔径; *v*. 镗孔,打眼

boring ['bɔ:riŋ] *n*. 镗孔,镗削加工

boring-mill 铣镗床

boron ['bɔ:rɔn] carbide 碳化硼

bottom line 底线,要点,关键之处

boundary lubrication 边界润滑

bracket ['brækit] *n*. 托架,轴承架,支座,括号

brake [breik] *n*. 制动器,刹车; *v*. 制动,减速,刹车

brass [brɑ:s] *n*. 黄铜

breakable ['breikəbl] *a*. 易破的,易碎的; *n*. 易破碎物

breakage ['breikidʒ] *n*. 破损,断裂,损坏

brittle ['britl] *a*. 脆性的,易碎的,易损坏的

broach [brəutʃ] *n*. 拉削,拉刀; *v*. 拉削

bronze [brɔnz] *n*. 青铜

browse [brauz] *v*.; *n*. 浏览,翻阅

built-up edge ['bilt'ʌpedʒ] *n*. 积屑瘤

bumper ['bʌmpə] *n*. 保险杠,缓冲器,阻尼器,减震器

buoyancy ['bɔiənsi] *n*. 浮力

burr [bə:] *n*. 毛刺

business performance 企业经营业绩

calculus ['kælkjuləs] *n*. 微积分,计算,演算

cam [kæm] *n*. 凸轮,偏心轮,样板,靠模,仿形板

camshaft ['kæmʃɑ:ft] *n*. 凸轮轴

cantilever ['kæntili:və] *n*. 悬臂; *v*. 使…伸出悬臂梁

capacitive [kə'pæsitiv] *a*. 电容的,电容性的

capacitor [kə'pæsitə] *n*. 电容器

carbide ['kɑːbaid] *n.* 碳化物,硬质合金
carbide tipped drill 硬质合金钻头
carbon steel 碳素钢,碳钢
carburize ['kɑːbjuraiz] *v.* 渗碳,碳化
carriage ['kæridʒ] *n.* 溜板
Cartesian coordinate 笛卡儿坐标,直角坐标
case-harden 表面淬火,表面渗碳硬化
cast iron 铸铁
categorize ['kætigəraiz] *v.* 分类,把…归类,区别
category ['kætigəri] *n.* 种类,类别,类型,等级
cavity ['kæviti] *n.* 空腔,孔穴,型腔,模腔
cement [si'ment] *n.* 胶结材料,粘合剂;*v.* 粘合,胶粘
center ['sentə] *n.* 中心,顶尖;*v.* 定心,对中
center distance 中心距,顶尖距,轴间距离
centerless ['sentəlis] *a.* 无心的,没有心轴的
centre of gravity 重心
centerpiece ['sentəpiːs] *n.* 主要特征,引人注目的东西
centrifugal [sen'trifjugəl] *a.* 离心的
ceramics [si'ræmiks] *n.* 陶瓷,陶瓷材料
chamfer ['tʃæmfə] *n.*;*v.* 在……开槽,倒棱,倒角,斜面
changeover ['tʃeindʒəuvə] *n.* 转换,改变,转向,调整,改装
chatter ['tʃætə] *n.*;*v.* 颤振,自激振动(即 self-excited vibration)
chilled cast iron 冷硬铸铁
chip [tʃip] *n.* 碎片,切屑,芯片;*v.* 切,削,刨
chip breaker 断屑槽,断屑前面
chronological [ˌkrɔnə'lɔdʒikəl] *a.* 按照年月顺序的
chuck [tʃʌk] *n.* 卡盘;*v.* 装夹,(用卡盘)卡紧
CIM = Computer Integrated Manufacturing 计算机集成制造
circular interpolation 圆弧插补
circumscribe ['səːkəmskraib] *v.* 与…外接,确定…的范围,限制
class of fit 配合级别,配合种类
classification [ˌklæsifi'keiʃən] *n.* 分类,归类,类别

cleanliness ['klenlininis] *n*. 清洁度
clearance ['kliərəns] *n*. 间隙,空隙
clearance fit 间隙配合
climb cutting 顺切,顺铣
clockwise ['klɔkwaiz] *a*. 顺时针方向的; *ad*. 顺时针方向地
closed kinematic chain 闭式运动链
closed-loop 闭环的
clutch [klʌtʃ] *v*.;*n*. 离合器
code number 代号,编码数
coded programme 编码的程序,用编码表示的程序
coding ['kəudiŋ] *n*. 编码,译码
cold forming 冷成型,冷态成型
cold work 冷加工,常温加工
cold-work ['kəuld'wə:k] *v*. 冷变形加工
collision [kə'liʒən] *n*. 碰撞,撞击,冲击,抵触
column ['kɔləm] *n*. 柱,柱状物,架,墩,(钻床等)立柱
commensurate [kə'menʃərit] *a*. 同等大小的,相应的,成适当比例的
competitive product 有竞争力的产品,拳头产品
compilation ['kɔmpi'leiʃən] *n*. 编辑,编码,汇编,编译程序
compliant [kəm'plaiənt] *a*. 应允的,依从的,顺从的,柔性的
composite ['kɔmpəzit] *a*. 合成的,复合的; *n*. 复合材料
compound rest (车床)小刀架
compressibility [kəmpresi'biliti] *n*. 可压缩性
compression [kəm'preʃən] *n*. 压力,压缩
concave ['kɔn'keiv] *a*.;*n*. 凹的,凹面型,凹面
concentration [,kɔnsen'treiʃən] *n*. 集中,浓缩,浓度,密度,集度
concentric [kɔn'sentrik] *a*.;*n*. 同心的,共轴的,集中的
concentricity [,kɔnsen'trisiti] *n*. 同心,同心度,集中
concomitant [kən'kɔmitənt] *a*. 伴随的,相伴的,随…而产生的
concrete ['kɔnkri:t] *n*. 混凝土
conductivity [,kɔndʌk'tiviti] *n*. 传导率,传导性
conduit ['kɔndit] *n*. 导管,水管,输送管

confidence level 置信水平
configuration [kənfigju′reiʃən] *n*. 形状,轮廓,构造形式,结构
conical [′kɔnikəl] *a*. 圆锥形的
connecting-rod 连杆
conservative [kən′sə:vətiv] *a*. 保守的,谨慎的,有余量的,守恒的
conserve [kən′sə:v] *v*. 保存,储备,节省,守恒
constrain [kən′strein] *v*. 强迫,强制,制约,约束,束缚
constraint [kən′streint] *n*. 限制,制约,约束,束缚
consumer electronics 消费电子产品
contaminant [kən′tæminənt] *n*. 沾染,杂质,污染物质,污染剂
contemplate [′kɔntempleit] *v*. 注视,思考,设想,预期,估计
contour [′kɔntuə] *n*. 轮廓,外形,周线,形状
contract out 订合同把工作包出去
control panel 控制面板
controlled experiment 核对实验,对照实验,控制性实验
convection [kən′vekʃən] *n*. 对流,迁移,传递
conveyor [kən′veiə] *n*. 输送设备,传送带
coolant [′ku:lənt] *n*. 冷却液,散热剂,切削液,乳化液
coordinate [kəu′ɔ:dinit] *n*. 坐标,坐标系,相同; *a*. 坐标的
coordinate-measuring machine 三坐标测量机
coplanar [kəu′pleinə] *a*. 共面的,同一平面的
core drill 扩孔钻
corrosion [kə′rəuʒən] *n*. 腐蚀,侵蚀,锈蚀
cottage industry 家庭手工业
cotter [′kɔtə] *n*. 栓,开口销,楔; *v*. 用销固定
counterbore [ˌkauntə′bɔ:] *v*. ; *n*. (平底)锪孔,锪沉头孔
conterclockwise [ˌkauntə′klɔkwaiz] *a*. ; *ad*. 逆时针方向的(地)
countersinking 锪锥孔
couple [′kʌpl] *n*. 力偶
coupling [′kʌpliŋ] *n*. 联轴器,连接,耦合
craft [krɑ:ft] *n*. 技能,工艺,行业; *v*. (用手工)精巧地制作
craftsman [′krɑ:ftsmən] *n*. 技工,工匠

craftsmanlike ['krɑːfsmənlaik] *a*. 展现手艺的,精巧的

crane [krein] *n*. 起重机,吊车,升降设备

crankshaft ['kræŋkʃɑːft] *n*. 曲轴,曲柄轴

creep failure 蠕变破坏

criteria [krai'tiəriə] *n*. criterion 的复数

criterion [krai'tiəriən] *n*. 标准,规范,准则,依据

critical speed 临界速度,临界转速

cross feed 横向进给

cross section 截面,横断面,剖面

cross slide 横刀架,横拖板

crosshatch ['krɔːshætʃ] *v*. 给…画交叉阴影线,给…画截面线,网纹线

cross-rail ['krɔːs'reil] *n*. 横导轨,横梁

cryogenic [ˌkraiəˈdʒenik] *a*. 冷冻的,低温的,制冷的,深冷的

crystalline ['kristəlain] *a*. 结晶的,晶状的; *n*. 结晶体,晶态

curl [kəːl] *n*. 卷,卷曲,扭曲; *v*. 卷曲,扭曲,成螺旋状

curvature ['kəːvətʃə] *n*. 弯曲,曲率,弧度

custom design 定制设计,用户设计,按用户需求设计

customize ['kʌstəmaiz] *v*. 定做,按规格改制

cutting edge 切削刃,刀刃

cutting load 切削载荷

cutting tool 刀具

cyanide ['saiənaid] *n*. 氰化物;用氰化物处理

cycle parameter 循环参数

cylinder ['silində] *n*. 圆柱,柱体,气缸,油缸

cylindrical coordinate system 柱面坐标系统

dampen ['dæmpən] *v*. = damp 阻尼,减振,缓冲,抑制,衰减

damper ['dæmpə] *n*. 阻尼器,减振器,缓冲器

damping ['dæmpiŋ] *n*.; *a*. 阻尼,减振,缓冲,衰减,抑制

dashpot ['dæʃpɔt] *n*. 减振器,缓冲器,阻尼器

debug [diːˈbʌg] *v*. 调整,调试,发现并排除故障,审查

deburr [di'bəː] *v*. 去毛刺,去飞翅

decarburization [diːˌkɑːbjuriˈzeiʃən] *n*. 脱碳作用

decimal point 小数点

decode [di:'kəud] v. 解译,译码,解码,译出指令

dedicated ['dedikeitid] a. 专用的

deflection [di'flekʃən] n. 偏差,偏移,弯曲,挠度

deformable [di'fɔ:məbl] a. 可变形的,应变的

deformation [,di:fɔ:'meiʃən] n. 变形,形变,扭曲,应变

degenerate [di'dʒenəreit] v. 退化,变质,简并

degrade [di'greid] v. 降低,降级,减低,降解

degree-of-freedom 自由度

depress [di'pres] v. 按下,压下

design change 设计变更

designated ['dezigneitid] a. 指定的,特指的

detail ['di:teil] n. 细节,详细;v. 详述,画细部图,细部设计

detail drawing 零件图

detailing 零件设计,绘工程图细节,细节设计

deterioration [ditiə'reiʃən] n. 变质,退化,恶化,变坏

detrimental [,detri'mentl] a. 有害的,不利的;n. 有害的东西

deviation [,di:vi'eiʃən] n. 偏离,偏移,差异,误差,偏差

diagram ['daiəgræm] n. 图表,简图,示意图;v. 用图表示出

die [dai] v. 模切,用模压成形;n. 模具,冲模,锻模

die casting 压力铸造

diecasting 压铸件,压铸法

differential [,difə'renʃəl] a. 有差别的,差动的;n. 差别,差动装置,差速器

difficult-to-machine material 难加工材料

dimensional [di'menʃənl] a. …维的,…度的,量纲的

diminish [di'miniʃ] v. 减少,减弱,缩小,削弱

dimple ['dimpl] n. 凹痕,坑,表面微凹,波纹

disassembly [,disə'sembli] n. 拆卸,拆除,拆开,分解,解体

displacement [dis'pleismənt] n. 位移

disregard [disri'ga:d] n.;v. 忽视,忽略

dissipation [,disi'peiʃən] n. 消散,浪费,消耗

distillation [disti'leiʃən] n. 蒸馏,蒸馏物,精华,精髓

distort [dis'tɔ:t] v. 扭曲,畸变,歪曲,把…弄得不正常

distortion [dis'tɔ:ʃən] n. 变形,扭曲,畸变

distributed intelligence 分布式智能

disturbance [dis'tə:bəns] n. 扰动,扰动量

dividing head 分度头

domain [də'mein] n. 领域,范围,区,界

downtime n. 停机时间,发生故障时间

draft [drɑ:ft] n. 草稿,草案,草图；v. 起草,设计

drafting ['drɑ:ftiŋ] n. 起草,绘图

drag [dræg] v. 拖,牵引,摩擦,拖着,阻碍；n. 阻力,摩擦力,阻尼

draughtsman ['drɑ:ftsmən] n. 制图员,绘图员,起草者

drawing ['drɔ:iŋ] n. 拉拔,拉延,拉制,冲压成形,图纸,图样

dresser 砂轮修整器

dressing ['dresiŋ] n. (砂轮的)修整

dressing diamond 金刚石修整工具,金刚石修整笔

drill 钻头

drill chuck 钻头卡盘

drill press 钻床

drilling head 钻床主轴箱

drive [draiv] n.；v. 传动,驱动,拖动,传动装置

drive plate 拨盘,也可写作 driver plate

ductile ['dʌktail] a. 延性的,可锻的,可塑的,易变形的

ductility [dʌk'tiliti] n. 延性,可锻性,韧性,可塑性

duplex ['dju:pleks] n.；a. 双(的),双重(的),双联的,二部分的

durability [,djuərə'biliti] n. 耐久性,持久性,耐用期限

durable ['djuərəbl] a. 耐用的,耐久的,坚固的；n. 耐久的物品

dynamics [dai'næmiks] n. 动力学,动力特性

ecological [ekə'lɔdʒikəl] a. 生态的,生态学的

elastohydrodynamics 弹性流体动力学

elastic [i'læstik] a. 弹性

electrical discharge machining (EDM)电火花加工

electrochemical [i'lektrəu'kemikl] a. 电化学的

elevated ['eliveitid] *a.* 高架的,升高的,提高的
elongation [ˌiːlɔŋ'geiʃən] *n.* 拉伸,伸长,延长,延伸率
encoder [in'kəudə] *n.* 编码器
end mill 立铣刀
end product 最后产物,最终结果,成品
endurance [in'djuərəns] *n.* 忍耐,持久(性),耐用度,耐疲劳强度,寿命
end-effector 末端执行器,末端操作器
engage [in'geidʒ] *v.* 啮合,切入
engagement [in'geidʒmənt] *n.* 啮合,吃刀量,切入
engine block 发动机缸体,发动机本体
engine lathe 普通车床,卧式车床
engineering drawing 工程图样,工程图
engineering practice 工程实践,技术实践
entangle [in'tæŋgl] *v.* 缠上,使…纠缠,卷入
epicyclic ['episaiklik] *a.* 周转圆的,外摆线的
epicyclic gear 行星齿轮
equilibrium [ˌiːkwi'libriəm] *n.* 平衡(状态,性,曲线)
erode [i'rəud] *v.* 侵蚀,蚀除
erosion [i'rəuʒən] *n.* 腐蚀,侵蚀,磨损,磨蚀
eruption [i'rʌpʃən] *n.* 喷出,爆发,萌出,喷出物
etch [etʃ] *v.* ; *n.* 蚀刻,腐蚀
exacerbate [eks'æsəːbeit] *vt.* 加重,使恶化,激怒
exacting [ig'zæktiŋ] *a.* 严格的,苛求的,艰难的,需付出极大努力的
excitation [ˌeksi'teiʃən] *n.* 刺激,扰动,干扰,激励,激振
expert system 专家系统
extension [iks'tenʃən] *n.* 伸长,延长
extrapolate ['ekstrəpəleit] *n.* 推断,外推,外插
extruding [eks'truːdiŋ] *n.* 挤压成形,压制,模压
fabrication [fæbri'keiʃən] *n.* 制造,生产,加工,装配
face milling 端面铣削,面铣
facility [fə'siliti] *n.* 设备,器材,实验室,工厂,机构
facing ['feisiŋ] *n.* 平面加工,端面加工,车平面

factory cost 生产成本,制造成本,工厂成本

far-flung 分布广的,范围广的,遥远的,漫长的

fastener ['fɑːsnə] n. 紧固件,连接件

fastening ['fɑːsniŋ] n. 连接,紧固件,连接物

feed motion 进给运动

feed per tooth 每齿进给量

feed rate 进给速度

feed rod 光杠

feed speed 进给速度

feedback ['fiːdbæk] n. 反馈

ferrous ['ferəs] a. 铁的,铁类的

ferrous material 钢铁材料,黑色金属

field experience 现场试验,现场经验

fillet ['filit] n. (内)圆角,倒角,齿根过渡曲面;v. 修圆,倒角

filter ['filtə] n. 过滤器,滤波;v. 过滤,滤波

filtering n. 过滤,滤除,滤波;a. 过滤的

fine adjustment 精密调整

finished part 成品零件,制成零件

finished product 成品

finishing ['finiʃiŋ] n. 精加工,最终加工,表面加工

finite element analysis (FEA) 有限元分析

fit [fit] v. 适合,适用,装配;n. 配合,装配,适合,符合,适当

fitter ['fitə] n. 装配工,修理工,钳工

fitting ['fitiŋ] n. 装配,匹配,配合,装修,配件;a. 适当的,相称的

fitting surface 配合面

fivefold ['faivfəuld] a. ad. 五倍(的),五重的

fixture ['fikstʃə] n. 夹具,夹紧装置

flaking ['fleikiŋ] n. 薄片,表面剥落,压碎;a. 易剥落的

flame hardening 火焰淬火

flange [flændʒ] n. 凸缘,法兰

flank [flæŋk] n. 侧面,后面,边,外侧

flank wear 后刀面磨损

flexible coupling 弹性联轴器,挠性联轴器
flood coolant system 液冷系统
flush [flʌʃ] v.;n. (强液体流)冲洗,洗涤,冲刷
flute [flu:t] n. 凹槽,(刀具的)排屑槽,容屑槽
follower ['fɔləuə] n. 从动轮,随动机构,从动件,推杆
forced vibration 强迫振动
forge [fɔ:dʒ] n.;v. 锻造,打制,锻工车间
forging ['fɔ:dʒiŋ] n.;a. 锻造(的),模锻,锻件
form grinding 成形磨削
formalize ['fɔ:məlaiz] v. 使成正式,使定型,形式化
forming ['fɔ:miŋ] n. 形成,构成,仿形,成形法,模锻,模铸,冲压,模压件
foundry ['faundri] n. 铸造厂,翻砂厂(车间)
fragment ['frægmənt] n. 碎片,碎屑,毛刺;v. 使成碎片,使分裂
fragmentation [ˌfrægmen'teiʃən] n. 碎裂,破裂,破碎作用,晶粒的碎化
frame [freim] n. 机架,固定构件,机身
framework ['freimwə:k] n. 框架,结构,体制
free vibration 自由振动
freehand ['fri:hænd] a. 徒手画的
free-body 自由体,隔离体
frictional ['frikʃənl] a. 摩擦的,由摩擦产生的
functional ['fʌŋkʃənl] a.;n. 功能的,起作用的,实用的
gage [geidʒ] n. 量具,测量仪表,标准,限度,范围
gash [gæʃ] n. 裂口,裂纹;v. 划开,造成深长切口
gauge length 标距(长度),计量长度
gear [giə] n. 齿轮,齿轮传动装置
gear hobbing 滚齿
gear shaper 插齿机
gear shaping 插齿
gearbox ['giəbɔks] n. 齿轮箱,变速箱
General Motors Corporation (美国)通用汽车公司
general plan 总体规划
generative ['dʒenərətiv] a. 能生产的,有生产力的,再生的,创成的

generative approach 创成法

geometric [ˌdʒiə'metrik] *a*. 几何的,几何图形的

geometric ratio 等比

geometrical [dʒiə'metrikəl] *a*. 几何的,几何图形的

glass-fiber reinforcement 玻璃纤维增强

glide [glaid] *n*.;*v*. 滑动,滑移

gradient ['greidiənt] *n*. 梯度,变化率

grain boundary 晶界,晶粒边界

graph [grɑ:f] *n*. 图表,曲线图;*v*. 用图表示

graphite ['græfait] *n*. 石墨;*v*. 涂上石墨

gravitation [ˌgrævi'teiʃən] *n*. 引力,重力,引力作用

grindability 磨削性,可磨性

grinder ['graində] *n*. 磨床

grinding ['graindiŋ] *n*.;*a* 磨削(的)

gripper ['gripə] *n*. 抓持器,夹持器,手爪

groove [gru:v] *n*. 槽,沟,空心槽

ground surface 磨削表面

groundwork ['graundwə:k] *n*. 基础工作,根据

group technology 成组技术

guidance ['gaidəns] *n*. 引导,制导,向导,引导装置

guideway *n*. 导轨

gun drill 枪钻,深孔钻

G-code G 代码

harden [hɑ:dən] *n*. 淬火,淬硬

hardness number 硬度值

hardwired *a*. 电路的,硬件实现的,硬连线的

harmonic ['hɑ:mɔnik] *n*. 谐波,谐波分量,谐振荡,谐函数

headstock ['hedstɔk] *n*. 头架,主轴箱,动力箱

heat conductivity 导热性,导热率

helical ['helikəl] *n*.;*a*. 螺线,螺旋线,螺旋状的

helical gear 斜齿轮

helical spring lockwasher 弹簧垫圈

heuristic [hjuə'ristik] *a.*; *n.* 启发式的,发展式的,渐进的,探索的
heuristic knowledge 启发式知识
hierarchy ['haiərɑːki] *n.* 体系,系统,层次,分级结构
high-cycle fatigue 高周疲劳
high-powered *a.* 大功率的,力量大的
hob [hɔb] *n.* 滚刀
hobbing ['hɔbiŋ] *n.* 滚削,滚齿
horizontal boring machine 卧式镗床
hose down 用水龙带冲洗,用软管洗涤
hot forming 热成形
household appliance 家用电器
hub [hʌb] *n.* 中心部分,衬套,轮毂
HV = Vickers hardness 维氏硬度
hydraulic [hai'drɔːlik] *a.* 水力的,液压的;*n.* 液压传动装置
hydraulic motor 液压马达
hydraulically [hai'drɔːlikəli] *ad.* 应用水力原理,液压地
hydrodynamic ['haidroudai'næmik] *a.* 流体的,流体动力(学)的
hydrostatic [ˌhaidrəu'stætik] *a.* 液体静力的,流体静力的
imperative [im'perətiv] *a.*; *n.* 命令的,强制的,必须履行的责任
imperfection [ˌimpə'fekʃən] *n.* 不完善,不完整,不足,缺点,缺陷
impetus ['impitəs] *n.* 原动力,动量,刺激,促进,推动
impingement [im'pindʒmənt] *n.* 碰撞,冲击,打击,侵入,冲突
implementation [ˌimplimen'teiʃən] *n.* 履行,实现,执行过程
impulse ['impʌls] *n.* 冲击,碰撞,脉冲,脉动
in mesh 齿轮互相啮合
inaccessible [ˌinæk'sesəbl] *a.* 不能接近的,不能进入的,进不去的,难接近的
incipient failure 早期故障,初期故障,将临故障
Inconel 因科合金,铬镍铁耐热耐腐蚀合金
indentation [ˌinden'teiʃən] *n.* 压痕,凹痕
indeterminate [ˌindi'təːminit] *a.* 不确定的,未定的,无法预先知道的
indexable insert 可转位刀片
indexable turning tool 可转位车刀

indexing ['indeksiŋ] n. 分度,转位,转换角度,换挡
induction hardening 高频淬火,感应淬火
induction [in'dʌkʃən] n. 引导,感应,电感
inductive [in'dʌktiv] a. 感应的,电感的
inertia [i'nə:ʃiə] n. 惯性,惯量,惰性
inertness [i'nə:tnis] n. 惰性,不活泼,无自动力
inextensible [ˌiniks'tensəbl] a. 不能扩张的,不能拉长的
infinitesimal [ˌinfini'tesiməl] a. 无穷小的,极微小的;n. 无穷小量
information age 信息时代
information highway 信息高速公路
ingredient [in'gri:djənt] n. 成分,要素
inhibit [in'hibit] v. 防止,阻止,抑制,防腐蚀
injection molding 注塑成型法,注射成型
installation [ˌinstə'leiʃən] n. 整套装置,设备,结构,安装
integrated circuit 集成电路
integration [ˌinti'greiʃən] n. 积分,集成,综合,整体化
integrity [in'tegriti] n. 完整性,完善,正直,诚实
intended function 预定功能,预期功能
interchangeability ['intəˌtʃeindʒə'biliti] n. 可交换性,互换性,可替代性
interdependence [ˌintədi'pendəns] n. 互相依赖,相关性
interdependent [ˌintədi'pendənt] a. 相互依赖,相互影响,相互关联
interdisciplinary 各学科之间的,边缘学科的,多种学科的
interface ['intəfeis] n. 界面,接口设备,连接装置,结合面
interfacial [ˌintə'feiʃəl] a. 分界面的,两表面间的,界面的
interfere [ˌintə'fiə] v. 干涉,干扰,妨碍,抵触,冲突,过盈
interference [ˌintə'fiərəns] n. 干涉,干扰,妨碍,过盈,相互影响
interference fit 过盈配合
intermittent [ˌintə'mitənt] a. 间歇的,间断的,断续的
interpolation [inˌtə:pəu'leiʃən] n. 插入,补入,插值法
interrelate [ˌintəri'leit] v. 相互有关,互相联系
interrelationship ['intəri'leiʃənʃip] n. 相互关系,相互联系,相互影响
interwind [ˌintə'waind] v. 互相盘绕,互卷

Intranet [intrə'net] n. 企业内部互联网

intrude [in'tru:d] v. 硬挤进,侵入,干涉,打扰,妨碍

intuition [ˌintu:'iʃən] n. 直觉,直觉的知识

inversion [in'və:ʃən] n. 颠倒,倒置,(四杆机构的)机架变换,变换

involute ['invəlu:t] n. 渐开线; a. 渐开的,错综复杂的

irreparably [i'repərəbli] ad. 不能修理地,不能弥补地,无可挽救地

irreversible [ˌiri'və:səbl] a. 不能撤回的,不能改变的

iteration [itə'reiʃən] n. 反复,重申,迭代,逐步逼近法

iterative ['itərətiv] a. 反复的,迭代的,重复的

jargon ['dʒɑ:gən] n. 行话,术语

jet [dʒet] n. 射流,水流; v. 喷出,喷射,射流

jig [dʒig] n. 夹具,夹紧装置

jigs and fixtures 夹具

jig borer 坐标镗床

jig boring machine 坐标镗床

jobbing ['dʒɔbiŋ] n. 做临时工,重复性很小的工作

jobbing shop 修理车间

job-shop production 单件、小批生产

joining ['dʒɔiniŋ] n. 连接,连接物

journal ['dʒə:nl] n. 轴颈,辊颈,枢轴

journal bearing [sli:v] n. 滑动轴承,向心滑动轴承

journeyman ['dʒə:nimæn] n. 熟练工人

junk [dʒʌŋk] n. 碎片,废物,废品; v. 丢掉,当作废物

key [ki:] n. 键

keyway 键槽

kinematic [ˌkaini'mætik] a. 运动的,运动学的

kinematics [ˌkaini'mætiks] n. 运动学

knee [ni:] n. (铣床的)升降台

knee-and-column type milling machine 升降台式铣床

knob [nɔb] n. 旋钮,圆形把手

lap [læp] n.; v. 研磨,研具,研盘

lapping ['læpiŋ] n. 研磨,精研

lathe [leið] *n.*; *v.* 车床,用车床加工,车削
lathe tool 车刀
law of gravitation 万有引力定律
lead time 前导时间,前置时间,产品设计至实际投产间的时间,订货至交货间的时间,交付周期
leadscrew ['liːdskruː] *n.* 丝杠
lever ['liːvə 或 'levə] *n.* 杠杆,操纵杆,手柄,把手
light-emitting diode(LED) 发光二极管
line shaft 主传动轴,动力轴
linear ['liniə] *a.* 直线的,线性的,一维的
lining ['lainiŋ] *n.* 衬层,涂层,覆盖
link [liŋk] *n.* 环,杆件,构件,机构中的运动单元体
linkage ['liŋkidʒ] *n.* 连杆机构,连接,低副运动链
lip [lip] *n.* 唇,凸出部分,刀刃,切削刃
liquid jet-cut 液体射流切割
list of materials 物料清单,材料清单
load-bearing *n.*; *a.* 承载(的),承重(的)
locating [ləu'keitiŋ] *n.* 工件定位,放样
locus ['ləukəs] (*pl.* loci) *n.* 轨迹,轨线,(空间)位置
longitudinal [ˌlɔndʒi'tjuːdinl] *a.* 长度的,纵向的,轴向的
longitudinal feed 纵向进给
longitudinally [ˌlɔndʒi'tjuːdinli] *ad.* 长度地,纵向地,轴向地
loosely fitting 松弛配合
lot size 批量,订购数量
low end 低级的,廉价的,低端的
lower and higher pairs 低副和高副
lower specification limit 尺寸下限,技术要求下限
lubricant ['ljuːbrikənt] *n.* 润滑剂,润滑材料; *a.* 润滑的
lubrication [ˌljuːbri'keiʃən] *n.* 润滑,润滑作用
machinability [məˌʃiːnə'biliti] *n.* 切削加工性,可加工性,机械加工性能
machine control unit 机床控制装置
machine tool 机床

machined surface 已加工表面
machinelike [mə'ʃi:nlaik] *a*. 像机器一样的,机器似的
machining center 加工中心
machining operation 机械加工,切削加工
machining process 加工过程
machining rate 加工速度
machinist [mə'ʃi:nist] *n*. 机械工人,机械操作者
magnesium [mæg'ni:ziəm] *n*. 镁
magnify ['mægnifai] *v*. 放大,扩大,增强
magnitude ['mægnitju:d] *n*. 大小,尺寸,量度,数值
maintenance ['meintinəns] *n*. 技术保养,维护,运转,操作
manipulator [mə'nipjuleitə] *n*. 机械手,操作机
manual operation 手工操作
manufacturability 工艺性,可制造性
manufacturing planning 制造计划,制造规程
manufacturing process 制造过程,制造方法
manufacturing step 工步
manuscript ['mænjuskript] *n*. 手稿,程序单,(零件的)加工图; *a*. 手抄的
margin [mɑ:'dʒin] *n*. 边缘,极限,刃带,刃边
market penetration 市场覆盖率,市场渗透
mass or bulk conserving 质量或体积不变(守恒)
mass production 大量生产,大批生产
material handling 物料搬运
mating surface 配合表面,啮合表面
MCU = machine control unit 机器控制装置
means [mi:nz] *n*. 方法,方式,工具,设备,装置
mechanical advantage 机械效益,机械增益
mechanical failure 机械故障
mechanics [mi'kæniks] *n*. 力学,机械学
mechanize ['mekənaiz] *v*. 实现机械化,在…之中使用机械
medium-duty 中型的,中等的,中批生产
meld [meld] *v*. = merge 融合,汇合,组合,配合,交汇

member 构件

memory chip 存储芯片

meshing ['meʃiŋ] *n*. 啮合,咬合,钩住

metallic [mi'tælik] *a*. 金属的,金属制的

metallurgical [ˌmetə'lə:dʒkəl] *a*. 冶金的,金相的

metalworking ['metəwə:kiŋ] *n*. 金属加工

methodology [meθə'dɔlədʒi] *n*. 方法学,方法论,方法手段

metric thread 公制螺纹

micron ['maikrɔn] *n*. 微米

microstrain [ˌmaikrəu'strein] *n*. 微应变

microstructure ['maikrəstrʌktʃə] *n*. 微观结构,显微组织

microswitch ['maikrəswitʃ] *n*. 微动开关,微型开关

microtome ['maikrətəum] *n*. 切片刀,(薄片)切片机

mill [mil] *n*. 工厂,轧钢厂,轧钢车间

milling ['miliŋ] *n*. 铣削

milling machine 铣床

minimize ['minimaiz] *v*. 使…成最小,最小化

minute [mai'nju:t] *a*. 微小的,微细的,精密的,细致的

misalignment ['misə'lainmənt] *n*. 未对准,不同轴性,不一致

miscellaneous function 辅助功能

mist coolant system 雾冷系统

mobile crane 移动式起重机

mobility [məu'biliti] *n*. 可动性,机动性,灵活性,活动度

mode of failure 故障形式,故障种类

moderate production 中批生产

modular ['mɔdjulə] *a*. 模数的,制成标准组件的,预制的,组合的

module ['mɔdju:l] *n*. 模数,模量,模件,组件,模块,可互换标准件

mold [məuld] = mould *n*. 模型,模具,模子

mold cavity 型模,阴模

molybdenum disulfide 二硫化钼

moment ['məumənt] *n*. 力矩

momentum [məu'mentəm] *n*. 动量

Monel ['məu'nel] *n*. 蒙乃尔铜镍合金

motoring ['məutəriŋ] *n*. 驾驶汽车;*a*. 汽车的

mounted ['mauntid] *a*. 安装好的,固定好的,安装在…上的

multifunctional *a*. 多功能的

multipurpose ['mʌlti'pəpəs] *a*. 通用的,多用途的,多功能的,多目标的

multiview 多视图,多视角

mutual attraction 相互吸引

nanometer ['nænə,mitə] *n*. 纳米

natural frequency 固有频率

natural vibration 自由振动,固有振动

necessitate [ni'sesiteit] *v*. 需要,使成为必要,以…为条件,迫使

needle bearing 滚针轴承

needlessly ['ni:dlisli] *ad*. 不需要地,无用地,多余地

nickel ['nikl] *n*. 镍

nitride ['naitraid] *n*.;*v*. 氮化,渗氮

nodular cast iron 球墨铸铁

nodular ['nɔdjulə] *a*. 节状的,球状的,团状的

noncircular ['nɔn'sə:kjulə] *a*. 非圆形的

noncoincident [,nɔnkəu'insidənt] *a*. 不重合的,不一致的,不符合的

noncrystalline *n*.;*a*. 非晶态的,非结晶的

nondestructive ['nɔndis'trʌktiv] *a*. 非破坏性的,无损的

nondimensional ['nɔndi'menʃənəl] *a*. 无量纲的

nonferrous ['nɔn'ferəs] *a*. 非铁的

nonferrous metal 有色金属

nonmetallic ['nɔnmi'tælik] *a*. 非金属的;*n*. 非金属物质

nonuniform ['nɔn'ju:nifɔ:m] *a*. 不均匀的,不一致的,非均质的

normal distribution 正态分布

novice ['nɔvis] *n*. 初学者,新手,生手

nylon ['nailən] *n*. 尼龙,尼龙织品

obscure [əb'skjuə] *v*. 模糊的,不清楚的,难解的

off the shelf 现成的,现用的

off-line 脱机,离线,指设备或装置不受中央处理机直接控制的情况

off-the-shelf 现成的,不用定做的,通用的
online ['ɔnlain] *a*. 联机的,与主机连在一起工作的,在线的
open loop 开环
operation sheet 工序卡片,操作卡片
operation speed 工作速度
operation 工序
operational load 工作负载,工作负荷
optimize ['ɔptimaiz] *v*. 优选,选择最佳条件,发挥最大作用
optimum ['ɔptiməm] *a*.;*n*. 最佳的,最佳状态,最适宜的
optional stop 任选停止,可选择停止
order of magnitude 数量级
orientation [ˌɔːrien'teiʃən] *n*. 定向,朝向,定位,方位
oscillate ['ɔsileit] *v*. 振荡,振摆,摇摆,游移
oscillatory ['ɔsileitəri] *a*. 振动的,振荡的,摆动的,摇动的
oscilloscope ['ɔsiləskəup] *n*. 示波器
outgrowth ['autgrəuθ] *n*. 长出,派生,支派,副产品,结果
outlet ['autlet] *n*. 排出口,流出口,排泄口,排水孔
out-of-balance 不平衡,失去平衡
overhead charge 企业一般管理费,经常费,间接费用
overload ['əuvələud] *v*.;*n*. (使)超载,超重,过负荷,使负担过重
override [ˌəuvə'raid] *v*. 超过,克服
overstress [ˌəuvə'stres] *n*.;*v*. 过载,超载,过度应力
packing ['pækiŋ] *n*. 包装,组装,填充物,密封垫,密封件
parallel ['pærəlel] *a*. 并行的,平行的,相同的;*n*. 平行线
parameter [pə'ræmitə] *n*. 参数,系数,特征值
part family 零件族,零件组
part list 零件表,明细表,材料清单
parting ['pɑːtiŋ] *a*. 分离的,离别的;*n*. 分离,切断
partition [pɑː'tiʃən] *n*.;*v*. 划分,区分,分割,分离
part-program 零件加工程序
passageway ['pæsidʒˌwei] *n*. 通道,通路
pattern ['pætən] *n*. 模型,图形;*v*. 模仿,仿造(after)

peening ['pi:niŋ] n. 用锤尖敲击,喷丸硬化,喷射(加工硬化法)
peripheral [pə'rifərəl] a. 周围的,外围的; n. 外部设备,辅助设备
peripheral speed 圆周线速度
permissible [pə'misəbl] a. 容许的,许可的,安全的
perpendicular [pə:pən'dikjulə] a. 垂直的; n. 垂直,正交,垂线
photodiode [,fəutəu'daiəud] n. 光敏二极管,光电二极管
piecemeal ['pi:smi:l] ad.; a. 逐点,逐渐,逐段,一部分一部分地
piezoelectric [pai'i:zəui'lektrik] a. 压电的
pin connected 铰接的,销接的
pinion ['pinjən] n. 小齿轮,传动齿轮
pinpoint ['pinpɔint] n. 针尖; a. 极精确的,细致的; vt. 准确定位,确认
piston ['pistən] n. 活塞,柱塞
pitch [pitʃ] n. 螺距,齿距,节距,俯仰
pitch circle 节圆
pitch diameter (螺纹)中径
planar ['pleinə] a. 平面的,在(同一)平面内的,二维的
planer ['pleinə] n. 龙门刨床
planer type milling machine 龙门式铣床
planing ['pleiniŋ] n.; a. 刨削(的)
planning sheet 工序单
plasmas ['plæzmə] n. 等离子
pneumatic [nju:'mætik] a. 气动的,气压的,充气的
polygonal [pɔ'ligənl] a. 多边形的,多角形的
porosity [pə:'rɔsiti] n. 多孔性,孔隙率,密集气孔
positioning [pə'ziʃəniŋ] n. 定位,位置控制
post processor 后置处理程序,后处理程序
powder-metallurgy bearing 粉末冶金轴承
predetermine ['pri:di'tə:min] v. 预定,注定,先对…规定方向
preexist ['pri:ig'zist] v. 先前存在
preheat ['pri:'hi:t] v. 预热,预先加热
preload ['pri:'ləud] n.; v. 预加荷载
preparatory code 准备功能代码

preprocess ['pri:'prəuses] v. 预先加上,预处理
preset [pri'set] v.;a. 预调,预先设置
pressurize ['preʃəraiz] v. (使)增压,对…加压,产生压力,使压入
prestress [pri:'stres] v.;n. 预加应力于,施加预应力
pretest ['pri:test] n.;v. 事先试验,预先检验
preventive maintenance 预防性维修,定期检修
primary motion 主运动
prime mover 原动机,原动力,牵引机,发动机
principle problem 首要问题
prism ['prizəm] n. 棱镜,棱柱
prism pair 移动副
process capability index 工序能力指数
process capability 工序能力,设备加工能力
process plan 生产工艺设计,工艺规程
process planner 工艺设计人员
process planning 工艺过程设计
process route 工艺路线
process sheet 工艺过程卡,工艺卡
process variation 加工偏差,加工变化范围
production line 生产线,流水线,装配线
production process 生产过程
production run 生产过程,生产运行
program stop 程序停止
programmable controller 可编程控制器
programmable logic controller 可编程逻辑控制器
programming 程序设计,程序编制
projection [prə'dʒekʃən] n. 投影,射影,预测,计划
proofing ['pru:fiŋ] n. 证明,验算,校对
proofread ['pru:fri:d] v. 校读,校定,校对
propeller [prə'pelə] n. 螺旋桨,推进器
proportional [prə'pɔ:ʃənl] a. (成正)比例的,平衡的,相称的
propulsion [prə'pʌlʃən] n. 推进,推进器

prototype ['prəutətaip] *n*. 原型,样机,模型机,样机
proximity [prɔk'simiti] *n*. 接近,贴近,近程
proximity sensor 接近传感器
public utility 公用事业,公共事业机构
pulley ['puli] *n*. 滑轮,皮带轮
pulsating ['pʌlseitiŋ] *a*.;*n*. 脉动(的),脉冲的,片断的
punch [pʌntʃ] *n*.;*v*. 打孔,穿孔
pyramid ['pirəmid] *n*. 棱锥,四面体;*v*. 成角锥形
qualitative ['kwɔlitətiv] *a*. 性质上的,定性的
quantitative ['kwɔntitətiv] *a*. 数量的,定量的
quenching ['kwentʃiŋ] *n*. 淬火
quill [kwil] *n*. 活动套筒,衬套,钻轴,空心轴
raceway 轴承座圈,滚道
rack [ræk] *n*. 齿条
radial ['reidjəl] *a*. 径向的
radial drill 摇臂钻床
radioactive ['reidiəu'æktiv] *a*. 放射性的,放射引起的
random vibration 随机振动
range sensor 距离传感器
rationale [ræʃiə'nɑːli] *n*. 基本原理,理论基础
rattle ['rætl] *v*. 发出喀啦声,发硬物震动声;*n*. 喀啦声
real time 实时,与发生的物理过程同步进行的计算
reaming ['riːmiŋ] *n*. 铰孔
recapitulate [riːkə'pitjuleit] *v*. 扼要重述,概括,重现,再演
reciprocating [ri'siprəkeitiŋ] *n*.;*a*. 往复(的),来回(的),交互
rectangular [rek'tæŋgjulə] *a*. 矩形的,直角的
recycle ['riːsaikl] *v*.;*n*. 再循环,回收,重复利用
redundant system 冗余系统
refinement [ri'fainmənt] *n*. 提纯,明确表达,改进的地方
reflectance [ri'flektəns] *n*. 反射,反射率,反射能力
reflex ['riːfleks] *n*. 反射,映像,反应能力
regime [rei'ʒiːm] *n*. 状况,状态,方式,方法

reliability [rilaiə'biliti] *n*. 可靠性,安全性
removal [ri'mu:vəl] *n*. 除去,切削,切除
removal rate 去除速度,体积加工速度
repeatability 可重复性,再现性,反复性
repetitious [ˌrepi'tiʃəs] *a*. 重复的,反复的
repetitive [ri'petitiv] *a*. 重复的
replication [ˌrepli'keiʃən] *n*. 重复,重现,仿作,复制过程
reproduction [ri:prə'dʌkʃən] *n*. 繁殖,再现,复制品
rerun [ri:'rʌn] *v*. 再开动,重新运转,重算
reside [ri'zaid] *v*. 驻留,居住,归于
residual stress field 残余应力场
resin ['rezin] *n*. 树脂,树脂制品;*v*. 用树脂处理
resistor [ri'zistə] *n*. 电阻器
resolution [rezə'lu:ʃən] *n*. 分辨率
resonance ['rezənəns] *n*. 共振,谐振
restate [ri:'steit] *v*. 重申,重新叙述
resultant [ri'zʌltənt] *a*. 组合的,总的;*n*. 合力,合成矢量,组合
retard [ri'tɑ:d] *v*.;*n*. 延迟,使停滞,阻滞,使减速
retract [ri'trækt] *v*. 缩回,缩进,收缩,取消
retrieve [ri'tri:v] *v*. 保持,恢复,检索
reversal [ri'və:səl] *n*. 颠倒,相反,换向,变号
reverse [ri'və:s] *v*.;*n*.;*a*.;*ad*. 颠倒(的),相反(的),改变方向,倒退
revolute joint 旋转关节
rework ['ri:'wə:k] *v*. 重作,再加工,重新加工后利用
rib [rib] *n*. 肋,加强肋;*v*. 加肋于,用肋状物加固
rivet ['rivit] *n*. 铆钉;*v*. 铆接,固定,钉铆钉
robotics [rəu'bɔtiks] *n*. 机器人学,机器人技术,机器人的应用
roller bearing 滚柱轴承
roller burnishing 滚压
rolling ['rəuliŋ] *n*. 滚轧,滚压,压延;*a*. 滚压的,辊轧的
rotary ['rəutəri] *a*. 旋转的,回转的
rotation [rəu'teiʃən] *n*. 旋转,转动

rough [rʌf] *a.* ; *ad.* 粗糙(的),不光的,粗(未)加工的; *v.* 粗制,粗加工

roundness ['raundnis] *n.* 圆度

round off 舍入,四舍五入

rpm = revolutions per minute 转数/分

rub [rʌb] *n.* ; *v.* 摩擦,磨损

runout ['rʌn'aut] *n.* 偏斜,径向跳动

rupture ['rʌptʃə] *v.* ; *n.* 破裂,断裂

salability [ˌseilə'biliti] *n.* 出售,销路,畅销

salient feature 特征

scalar ['skeilə] *n.* ; *a.* 数量(的),标量(的)

scanner ['skænə] *n.* 扫描器,扫描仪,多点测量仪

schedule ['skedjuːl] *n.* 时间表,进度表,计划表,进程,预订计划

scrap [skræp] *n.* 碎片,切屑,废品,废渣; *a.* 碎片的; *v.* 废弃,使成碎屑

scraping ['skreipiŋ] *n.* 刮削

screw [skruː] *n.* 螺旋丝杆,螺钉

screwdriver 螺丝刀,螺丝起子,改锥

screw fastener 螺丝紧固件

scrub [skrʌb] *v.* 擦洗,洗涤,洗刷; *n.* 擦洗,磨

seal [siːl] *n.* 封口,密封装置,垫圈

sealing [siːliŋ] *n.* 密封,封接

seizure ['siːʒə] *n.* 轧住,咬住,卡住,塞住,咬缸

self-align *v.* ; *n.* 自动调整,自位,调心,自行对准

self-aligning bearing 调心轴承,自位轴承,自动定心轴承

self-locking [self'lɔkiŋ] *a.* 自锁的

sense [sens] *n.* ; *v.* 检测,显示,方向

separator ['sepəreitə] *n.* 分离,隔离,轴承保持架

sequentially [si'kwenʃəli] *ad.* 顺序地

service conditions 操作条件,使用条件,工作情况

service performance 使用性能

serviceability [ˌsəːvisə'biliti] *n.* 适用性,操作性能,耐用性,可维修性

servo ['səːvəu] *n.* 伺服机构,伺服电机,伺服传动装置

servomotor ['səːvəuˌməutə] *n.* 伺服电动机

set up 装夹,安装

setting ['setiŋ] n. 安置,设置,调整,设定,机械部件等的工作位置

setup ['setʌp] n. 装夹(工件在机床上或夹具中定位、夹紧的过程),安装(工件或装配单元经一次装夹后所完成的那一部分工序),编排(在一个数控循环操作之前,确定用于控制和显示的一系列功能)

shaft [ʃɑ:ft] n. 轴

shaft bearing 轴承

shaper ['ʃeipə] n. 牛头刨床

shaping ['ʃeipiŋ] v. 刨削,压力加工;a. 成形的

shaving ['ʃeiviŋ] n. 剃齿,剃齿法

shear [ʃiə] v. 剪切,切断;n. 剪切,剪力,剪应变

shear force diagram 剪力图

sheave [ʃi:v] n. 带轮(grooved pulley),V 带轮

shield [ʃi:ld] n. 防护屏,挡板,防尘板;v. 防护,起保护作用

shock [ʃɔk] n. 冲击,冲撞,打击

shock-excitation n. 震激,冲击激励

shoddy ['ʃɔdi] a. 以次充好的,劣质的

shop floor 车间,工厂里的生产区,生产区的工人

shorthand ['ʃɔ:thænd] n. 简略的表示方法

shorthand notation 简化符号

shrinkage ['ʃriŋkidʒ] n. 收缩,收缩率,减少

shutdown ['ʃʌtdaun] n. 关闭,断路,停工,停止

sideward ['saidwəd] a.;ad. 侧面(的),向旁面的,从旁边的

side-effect n. 副作用,边界效应

silicon wafer 硅片

single-point tool 单刃刀具,单点刀具

single-threaded screw 单头螺纹螺钉

sign convention 符号规定

sinusoidal [sainə'sɔidəl] a. 正弦波的,正弦曲线的

sketch [sketʃ] n. 草图,简图,设计图,图样设计;v. 画简图

slab milling 平面铣削,也可写作 peripheral milling

slider-crank mechanism 曲柄滑块机构

slip [slip] v.; n. 滑动,滑移,润滑性,滑动量,逃逸

slurry ['slə:ri] n. 悬浮液,膏剂,软膏,磨料粉浆

smear [smiə] v. 涂,抹,弄脏,弄污,使…轮廓不清

socket ['sɔkit] n. 插座,插口,套筒

socket wrench 套筒扳手

solid modeling 实体造型

solidification [sə,lidifi'keiʃən] n. 凝固(作用),固化

sound [saund] n. 声音,a. 完整的,正确的,合理的,有根据的

space coordinate system 空间坐标系

spall [spɔ:l] v. 削,割,打碎,剥落,脱皮; n. 裂片,碎片

special-purpose 专用的,单一用途的,特殊用途的

specfic gravity 比重

specific heat 比热

specification [,spesifi'keiʃən] n. 详细说明,尺寸规格,技术要求,技术参数

specification width 规定的尺寸范围,技术要求范围

specify ['spesifai] v. 规定,确定,拟订技术条件,表示…的规格

specimen ['spesimin] n. 样品,样本,试样,试件

speed reducer 减速器

spindle ['spindl] n. 心轴,主轴,转轴

spline [splain] v. 把…刻出键槽,用花键连接; n. 花键,键槽,样条函数

spot facing 锪端面

spring [spriŋ] n. 弹簧,发条; v. 跳跃,弹出

springback ['spriŋbæk] n. 回跳,回弹,弹回,弹性后效

spur [spə:] n. 齿,正齿,支承物,刺激,鼓励,推动

spur gear 直齿轮

squeeze [skwi:z] v. 挤压,压缩; n. 压榨,挤压

squeeze out 挤压,压出

staggering ['stægəriŋ] n. 交错; a. 交错的,惊人的,压倒的

stainless steel 不锈钢

stamp [stæmp] n. 压制,捣碎; v. 冲压成形,模锻

standard deviation 标准偏差,标准差,均方差

standard process 典型工艺

standardization [stændədai′zeiʃən] *n*. 标准化,规格化,标定,校准
statically indeterminate 静不定的,超静定的
statics [′stætiks] *n*. 静力学,静止状态
static-equilibrium position 静平衡位置
stationary [′steiʃnəri] *a*. 静止的,固定的,平稳的;*n*. 固定物
statistical [stə′tistikəl] *a*. 统计的,统计学的
statistically [stə′tistikəli] *ad*. 统计学地,统计地
steady-state vibration 稳态振动
steam turbine 汽轮机
stem [stem] *n*. 杆,棒,柄,柱,轴
stem from 由…产生的,产生于,起源于,出身于
stepped shaft 阶梯轴
stepper motor 步进电机
stiffness [′stifnis] *n*. 刚性,刚度,稳定性
strain [strein] *n*. 应变
strain gauge 应变仪,应变传感器,应变片
strength [strenθ] *n*. 强度
strength of materials 材料力学,也可以写作 mechanics of materials
stress [stres] *n*. 应力,受力状态,重点;*v*. 强调,着重,加压力
stress concentration 应力集中
stress corrosion 应力腐蚀
stress corrosion cracking 应力腐蚀裂纹
stroke [strəuk] *n*. 冲程,行程
strut [strʌt] *n*. 支柱,支杆,竖直构件支柱
sturdy [′stə:di] *a*. 坚固的,结实的,加强的
subassembly [′sʌbə′sembli] *n*. 组件,部件,局部装配,组件装配
subcontractor [′sʌbkən′træktə] *n*. 第二次转包的工厂,转包人,小承包商
subdivide [′sʌbdi′vaid] *v*. 细分,再分,重分;*n*. 分水岭
subsidiary [səb′sidjəri] *a*. 辅助的,次要的,附属的;*n*. 附属机械
subsystem [′sʌbˌsistim] *n*. 子系统
succinct [sək′siŋkt] *a*. 简洁的
suitability [ˌsju:tə′biliti] *n*. 合适,适当,适宜性

superalloy [sjuːpəˈæloi] n. 超耐热合金,高温合金
superfinishing n. 超精加工
superimpose [ˈsjuːpərimˈpəuz] v. 重叠,叠加,附加,把…放在…上面
supersede [ˌsjuːpəˈsiːd] v. 代替,接替,取代,置换,废除
suppress [səˈpres] n. 压制,扑灭,抑制,制止,排除,隐蔽
surface roughness 表面粗糙度
surface texture 表面结构(是表面粗糙度,表面波纹度,表面纹理和表面缺陷等的总称)
suspension [səsˈpenʃən] n. 悬挂,悬置,悬挂装置
swarf [swɔːf] n. 细铁屑
swing [swiŋ] v.; n. 摇摆,摆动,最大回转直径,车床床面上最大加工直径
swivel [ˈswivl] n. 旋转轴承,转体; v. 旋转,用活节连接
synchronize [ˈsiŋkrənaiz] v. 使同步,同时进行,同时发生
synthesize [ˈsinθisaiz] v. 合成,综合,接合
table of contents 目录
tabular [ˈtæbjulə] a. 表格式的,列表的; n. 表格,表值
tabulate [ˈtæbjuleit] v. 把…制成表,用表格表示,精简,概括
tackle [ˈtækl] v. (着手)处理,从事,对付,解决
tactile [ˈtæktail] a. 触觉的,能触知的
tactile sensor 触觉传感器
tailstock [ˈteilstɔk] n. 尾座,尾架
tang [tæŋ] n. 柄脚,柄舌,扁尾
tangentially [tænˈdʒənʃəli] ad. 成切线地
tangible [ˈtændʒəbl] a. 有形的,真实的,实质的,明确的
tap [tæp] n. 丝锥
taper [ˈteipə] n. 圆锥,锥体
tapered [ˈteipəd] a. 锥形的,斜的
tapping [ˈtæpiŋ] n. 攻丝,攻螺纹
teethe troubles 事情开始时的暂时困难
temper [ˈtempə] v.; n. 回火
template [ˈtemplit] n. 样板,型板; v. 放样
tension [ˈtenʃən] n. 张力,拉力,张开,拉伸; v. 拉伸,拉紧

terminology [tə:mi'nolədʒi] *n*. 专门名词,术语
thermal ['θə:məl] *a*. 热的,热力的
thermal fatigue 热疲劳
thervnal conductivity 导热率
thermal expension 热膨胀
thermodynamics ['θə:məudai'næmiks] *n*. 热力学
thermoplastic ['θə:məu'plæstik] *n*.; *a*. 热塑性,塑性,热塑性的
threaded fastener 螺纹紧固件,螺纹联接件
titanium [tai'teiniəm] *n*. 钛
through-harden *v*. 整体淬火,淬透
through-hardening *n*. 全部硬化,整体淬火,淬透
through hole 通孔
thrust [θrʌst] *v*. 推入,塞,把…强加于; *n*. 推力,轴向力
thrust washer 止推垫圈,止推环
tightness ['taitnis] *n*. 紧密性,松紧度,密封性
tilted ['tiltid] *a*. 倾斜的,与…成角度的
timing ['taimiŋ] *n*. 定时,计时,时间控制,调整,校准
titanium [tai'teiniəm] *n*. 钛
tolerance ['tolərəns] *n*. 公差,允许限度;给(机器零件等)规定公差
tolerance grade 公差等级
tolerance zone 公差带
tool changer 换刀装置
tool cutting edge angle 主偏角
tool number 刀具号
tool post 刀架
tooling ['tu:liŋ] *n*. 工装,工艺装备(产品制造过程中使用的各种工具的总称,包括刀具、量具、夹具等)
top view 俯视图
torn [tɔ:n] *a*. 不平的(表面),有划痕的
torque ['tɔ:k] *n*. 转矩,扭矩
torsion ['tɔ:ʃən] *n*. 扭转
torsional ['tɔ:ʃənl] *a*. 扭转的,扭力的

torsional-vibration damper 扭振阻尼器
total quality management 全面质量管理
touch sensor 接触传感器
triplicate ['triplikit] *a.* 三倍的; *v.* 使增至三倍; *n.* 三个相同物中的第三个
tracer ['treisə] *n.* 追踪装置, 随动装置, 仿形板
tractable ['træktəbl] *a.* 易处理的, 易加工的
trade jargon 本专业的行话
tradeoff 折衷(方法, 方案), 权衡, 协调
train [trein] *n.* 系列, 链, 将不同装置串联起来的系统
transducer [træns'dju:sə] *n.* 变换器, 转换器, 传感器
transient ['trænziənt] *a.* 瞬态的, 暂时的, 不稳定的
transition fit 过渡配合
translate [træns'leit] *v.* 移动, 平移, 作直线运动, 转化
translation [træns'leiʃən] *n.* 平移, 位移, 平行位移, 直线运动, 转化
transmission [trænz'miʃən] *n.* 传动装置, 变速箱
transmitter [trænz'mitə] *n.* 发送器, 传递器, 发射机
traverse ['trævə:s] *v.* 横移, 通
tread [tred] *n.* 踩, 踏, 滑动面, 轮胎花纹
tremor ['tremə] *n.* 振动, 颤抖, 地震
trepan [tri'pæn] *n.* 环钻, 套料钻; *v.* 套孔, 从…中取出岩心
trepanning drill 环孔钻, 套料钻
trial ['traiəl] *n.* 试验, 试算, 检查; *a.* 试验性的, 尝试的
trial and error 试算法, 试配法, 反复试验, 不断摸索
triaxial [trai'æksiəl] *a.* 三维的, 三轴的, 空间的
tribology [trai'bɔlədʒi] *n.* 摩擦学
trigonometric [ˌtrigənə'metrik] *a.* 三角学的, 三角的
trimming ['trimiŋ] *n.* 整形, 修整
triplex ['tripleks] *n.*; *a.* 三个部分(的), 由三个组成(的), 三联(的)
true [tru:] *a.* 真正的; *v.* 调整, 精密修整, 砂轮的整形(产生确定的几何形状)
turbine ['tə:bin] *n.* 涡轮机, 透平机, 汽轮机, 水轮机
turnaround 转变, 转向, 突然好转
turnaround time 解题周期, 周转时间

turning ['tə:niŋ] n. 旋转,车削,切削外圆

turning tool 车刀

turntable ['tə:nteibl] n. 转台,回转台,回转机构

turret ['tʌrit] n. (机床刀具)转塔,六角刀架

turret lathe 转塔车床

twist [twist] v. 使扭转,扭,使转动

twist drill 麻花钻头

types of production 生产类型

ultrasonic machining 超声加工,超声波加工

uncertainty ['ʌn'sə:tnti] n. 不定因素,不确定性,不清楚

uncommitted ['ʌnkə'mitid] a. 自由的,不受约束的,不负义务的,独立的

unconditional ['ʌnkən'diʃənl] a. 无条件的,无限制的

unconstrained ['ʌnkən'streind] a. 不受约束的,自由的,非强迫的

unconventional ['ʌnkən'venʃnl] a. 非传统的,非常规的,不一般的

undefined ['ʌndi'faind] a. 未规定的,不明确的,模糊的

undo ['ʌn'du:] (undid, undone) v. 拆开,松开,使恢复原状,取消

unity [ju:niti] n. 单一,一,唯一,统一,单位

unqualified ['ʌn'kwɔlifaid] a. 不合格的,无条件的,绝对的,彻底的

unquenchable [ʌn'kwentʃəbl] a. 不能熄灭的,止不住的,不能遏制的

unsettled [ʌn'setld] a. 不稳定的,不安定的,未解决的,混乱的

unsophisticated ['ʌnsə'fistikeitid] a. 不复杂的,简单的

unused ['ʌn'ju:zd] a. 不用的,未利用的,新的,不习惯的

unzip ['ʌn'zip] v. 拉开(拉链)

update [ʌp'deit] v. 使…现代化,适时修正,不断改进,革新

upgrade ['ʌp'greid] v.; n. 提高等级,提高标准,提升,改进

upper specification limit 尺寸上限,技术要求上限

upright ['ʌprait] a.; ad. 笔直的,竖立的, n. 支柱

upright drilling machine 立式钻床

up-close 在很近距离内的,因近距离的仔细观察而显示或提供的详细报道

vacuum cup 真空吸盘

variable ['vɛəriəbl] a. 变化的,易变的; n. 变量的,变数,参数,易变的东西

variant ['vɛəriənt] a. 不同的,差异的,变量的,多样的; n. 派生,变样

variant approach 派生法

vector ['vektə] *n*. 矢量,向量

vee [vi:] *n*.; *a*. V 字形(的),V 型(的)

vendor ['vendɔ:] *n*. 卖主,供货商

versus ['və:səs] *prep*. …对,与…比较,…与…的关系曲线,作为…的函数

vertical boring machine 立式镗床

vertical tracer lathe 立式仿形车床

vibration [vai'breiʃən] *n*. 振动

virtual company 虚拟公司

viscous ['viskəs] *a*. 粘性的,粘稠的

vise [vais] *n*. 虎钳,台钳; *v*. 钳住,夹紧

vitrified bond 陶瓷结合剂

vitrify ['vitrifai] *v*. (使)玻璃化,使成玻璃状物质

volcanic [vɔl'kænik] *a*. 火山的,由火山作用所引起的

washer ['wɔʃə] *n*. 垫圈

way 导轨

wear [wɛə] *v*.; *n*. 磨损,磨蚀,消耗,耗损

wedge [wedʒ] *n*. 楔; *v*. 楔入,楔牢

wedge action 楔紧作用

wedge-shaped zone 楔形区

welding ['weldiŋ] *n*.; *a*. 焊接(的),熔接(的),焊缝

weldment ['weldmənt] *n*. 焊件

well balanced 各方面协调的,匀称的,平衡的

well-deserved 理所当然的,当之无愧的

well-established 固定下来的,得到确认的,良好的

what-if 假设分析,作假定推测

wheel-and-axle 轮轴

work envelope 工作包络,加工包迹

workhold ['wə:khəuld] *v*. 工件夹持

workmanship [wə:kmənʃip] *n*. 手艺,作工,工作质量

workpiece ['wə:kpi:s] *n*. 工件

work-in-process 在制品

worm 蜗杆
worm gear 蜗轮,蜗轮装置
worm gear speed reducer 蜗轮减速器
worm wheel 蜗轮
wrench [rentʃ] *n.*;*v.* 扳手,拧,扳紧,扭转
wrist pin 活塞销,曲柄销,偏心轴销
yield failure 屈服破坏
zero defect 无缺陷
zero line 基准线,零线(公差与配合中确定偏差的一条基准直线)
zero reference point 零参考点,基准零点
zinc [ziŋk] *n.* 锌
zinc-base alloy 锌基合金